Be prepared...
To learn...
To succeed...

REA's preparation for the Florida Geometry EOC is fully aligned with the Next Generation Sunshine State Standards and Common Core State Standards.

Visit us at ***www.rea.com***

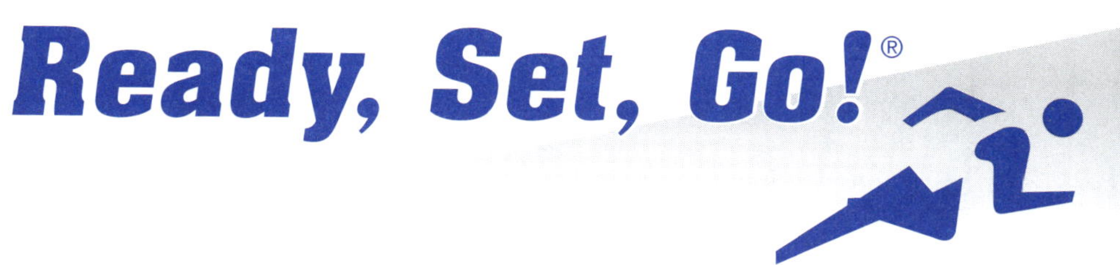

FLORIDA
End-of-Course Assessment
Geometry

Rebecca Dayton
Mathematics Curriculum Specialist

With a foreword by
Michael Coon
Anclote High School
Holiday, Florida

The benchmarks presented in this book were created and implemented by the Florida Department of Education (FLDOE).
For further information, visit the FLDOE website at *http://fcat.fldoe.org*.

Research & Education Association
61 Ethel Road West
Piscataway, New Jersey 08854
E-mail: info@rea.com

Ready, Set, Go!®
Florida Geometry End-of-Course Assessment

Printed in the United States of America

Library of Congress Control Number 2012955614

ISBN-13: 978-0-7386-1112-9
ISBN-10: 0-7386-1112-3

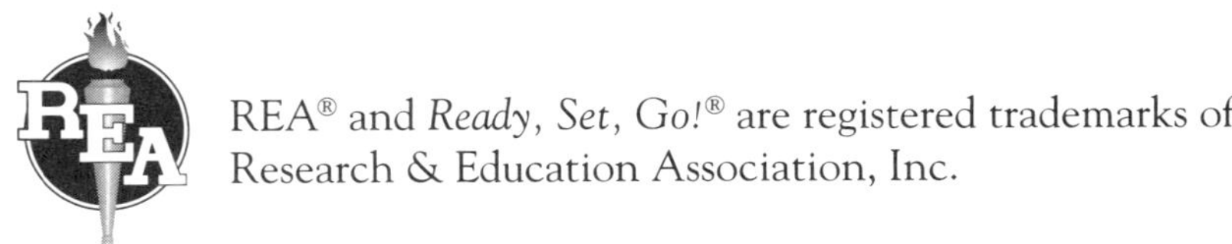

Contents

Practice Tests A & B are *also* available online at *www.rea.com/studycenter*

Foreword

With the arrival of high-stakes end-of-course assessments (EOCs) in Florida's high schools, students and teachers need support to prepare for success. Although teachers are given standards, we are left with questions about how those standards will be assessed on the EOCs. Some standards will point to a broad understanding, while other standards will concern a specific calculation. In facing these standards-driven complexities, students and teachers can often be left with more questions than answers about the EOC.

Rebecca Dayton's *Florida Geometry End-of-Course Assessment* provides the clarity and clarification that geometry students and teachers need for successful completion of the EOC. The chapters are broken down by topic, and each topic contains clear examples while indicating the standards that will be assessed in the topic. Every standard in this book is assessed either directly or indirectly on the EOC and fully aligned with the Common Core State Standards.

Ms. Dayton has "unpacked" and rated each standard as having low, medium, or high difficulty, so any reader will know the complexity of the concept being explained immediately. The examples in the chapters are conceptually aligned and contain some of the clearest explanations I've seen. At the end of the chapters, the reader will be able to take a quiz that assesses understanding of the concepts in the chapter. These assessments will help track the success rate of the concepts and will ultimately improve a student's score on the EOC. The book concludes with two practice tests aligned with the official geometry EOC. The two full-length practice tests will reveal the student's mastery of concepts while indicating which concepts still need to be reviewed.

Whether you are a geometry student or teacher, this book will give you a leg up on the EOC. I'm convinced that successful reading and completion of the activities in this book will raise students' scores.

Michael Coon
Anclote High School
Holiday, Florida
January 2013

Michael Coon teaches mathematics for college readiness at Anclote High School. He is also actively involved in Common Core curriculum planning.

About Our Author

Rebecca Dayton is currently a Mathematics Curriculum Specialist for the Piscataway (N.J.) Township Schools, a position she has held for the past five years. Ms. Dayton earned her B.A. in mathematics from Rutgers University. She taught high school mathematics for seven years.

About Our Technical Editor

Mary Berlinghieri-Willi is a longtime geometry teacher, math editor, and tutor to middle school, high school, and college math students.

About Research & Education Association

Founded in 1959, Research & Education Association (REA) is dedicated to publishing the finest and most effective educational materials—including study guides and test preps—for students in elementary school, middle school, high school, college, graduate school, and beyond.

Today, REA's wide-ranging catalog is a leading resource for teachers, students, and professionals. Visit *www.rea.com* for a complete listing of all our titles.

Acknowledgments

In addition to our author and technical editor, we would like to thank Larry B. Kling, Vice President, Editorial, for his overall guidance, which brought this publication to completion; Pam Weston, Publisher, for setting the quality standards for production integrity and managing the publication to completion; Alice Leonard, Senior Editor, for project management; Mel Friedman, Lead Math Editor, for project management and for quality control of the math content; and Kathy Caratozzolo of Caragraphics for typesetting the manuscript.

Introduction

Passing the Florida Geometry EOC Test

About This Book

This book, along with REA's true-to-format practice tests, provides you with the most up-to-date preparation for the Florida Geometry End-of-Course Assessment. Known simply as the Geometry EOC, this computer-based test measures your mastery of geometry.

This test prep gives you all the review and practice you need to succeed on exam day. By studying the review and taking our practice tests (either in the book or online), you can pinpoint what you already know and focus on the areas where you need to spend more time studying.

Each of our nine review chapters covers one of the major Geometry EOC subject areas and shows you how each area will be tested. Easy-to-follow examples and a step-by-step approach help you build your knowledge and confidence as you study.

Our two full-length practice tests give you the most complete picture of your strengths and weaknesses. After you've finished reviewing with the book, take our practice exams online at the REA Study Center (*www.rea.com/studycenter*). Because the Geometry EOC is computer-based, we recommend that you take the practice tests online to simulate test-day conditions. Each online test gives you instant score reports, diagnostic feedback, and onscreen detailed answer explanations.

If you're studying and don't have Internet access, you can take the printed tests in the book. These are the same tests offered at the REA Study Center, but without the added benefits of timed testing conditions and diagnostic score reports.

About the Test

The Florida EOC Assessments are part of Florida's Next Generation Strategic Plan, which is designed to increase student achievement and improve college and career readiness. EOC assessments are computer-based, criterion-referenced assessments that measure specific standards that have been developed for several courses, including Algebra 1, Biology 1, Geometry, and U.S. History.

How is this test given?

For the majority of students, the EOC assessments are computer-based. Exceptions are made for students with disabilities who need to take EOC assessments on paper.

What is the format of the test?

The Geometry EOC assessment is delivered in multiple forms with a maximum of 65 questions that include 35–40 multiple-choice items (MC) and 20–25 fill-in response items (FR). Some questions will be field-test items and are *not* included in student scores. You will also have a reference sheet that you can consult. According to the state's specifications, multiple-choice questions should take an average of one minute to answer. You should allocate about one and one-half minutes per fill-in question. You will be allowed to use a calculator on this exam.

The assessment will be given in one 160-minute session with a 10-minute break after the first 80 minutes.

Who takes this test?

All students enrolled in and completing the following courses take the Geometry EOC:

- Geometry–1206310
- Geometry Honors–1206320
- 1B Middle Years Program Geometry Honors–1206810
- Pre-AICE Mathematics 2–1209820

Preparing for Computerized Testing

Official Practice Tests

Prior to taking the Geometry EOC Assessment, students are required to participate in a practice-test session at their school in order to become familiar with the testing tools and platform. These are computer-based practice tests, called Electronic Practice Assessment Tools (ePATs). The online computerized practice tests included with this book will give students a valuable head start.

E-Tools for Exam Day

The tools and resources available to you on test day will vary slightly depending on the subject area assessed. All students taking a computer-based assessment will have access to the following e-tools in the computer-based platform:

- **Review:** Students use this e-tool to mark questions to be reviewed at a later time. Before exiting the assessment and submitting their responses, students are taken to a screen that identifies questions that are answered, unanswered, and marked for review.
- **Eliminate Choice:** Students use this tool to mark answer choices that they wish to eliminate.
- **Highlighter:** Students use this tool to highlight sections of the question or passage.
- **Eraser:** Students use the eraser to remove marks made by the highlighter or the eliminate-choice tool.
- **Help:** Students may click the Help icon to learn more about the e-tools. The Help text appears in a separate window.
- **Calculator:** Students are provided access to a scientific calculator, which appears in a pop-up window. They may request the use of a hand-held scientific calculator if they are uncomfortable with the online version.
- **Exhibit:** Students are provided a reference sheet of commonly used formulas and conversions to work the test questions. The reference sheet appears in a pop-up window under the exhibit icon.

- **Directions:** Students are also provided directions for completing fill-in response questions and a diagram and helpful hints for the appropriate calculator under the exhibit icon.

In addition, students are provided with a four-page hard-copy work folder to use as scratch paper. These will be collected and either securely stored or destroyed.

What to Do Before the Test

- **Pay attention in class.**
- **Carefully work through the chapters and problems in this book.** Mark any topics that you find difficult and rework them.
- **Take the practice tests and become familiar with the format of the Geometry EOC Assessment.** Do everything possible to take these tests under simulated conditions—time yourself, stay calm, and pace yourself. As we said earlier, multiple-choice questions should take about one minute and the fill-in questions about one and one-half minutes. We make this easy with our online practice tests, which provide a timed, auto-scored experience. (*www.rea.com/studycenter*)

What to Do During the Test

- **Read the questions carefully to make sure you understand what is being asked of you.** Every word in the question gets you that much closer to the answer.
- **Read all of the possible answers.** Even if you think you have found the correct response, do not automatically assume that it is the best answer. Read through each answer choice to be sure you are not jumping to conclusions.
- **Use the process of elimination in multiple-choice questions.** Process of elimination is one of the best techniques in solving these types of questions. Try to eliminate those choices that appear obviously incorrect. For each one you can eliminate, you've dramatically increased your odds of answering correctly.

If you eliminate two choices, for example, you now have a 50% chance of answering the question correctly.

- **Work on the easier questions first.** If you find yourself working too long on one question, move on to the next question. When you've reached the end of the test, there will be a window on the computer that will pop up and tell you which questions were unanswered so you can go back to them.
- **Be aware of the correct units.** If the question asks for *meters*, make sure you're not selecting an answer choice with a number that looks correct, but is in *feet*.
- **In fill-in response questions, make sure your answer is complete.** Again, check the types of units of measurement. For example, is the test asking for square inches or cubic inches?
- **Answer all of the questions.** You will not be penalized for incorrect answers, so you'll increase your chances of improving your score by guessing. Even one "good guess" can increase your score by a point.
- ***Relax.***

Good luck!

Overview of the Geometry EOC Assessment

This test is aligned with Florida's Next Generation Sunshine State Standards and with the Common Core State Standards. Florida's standards are presented below.

Next Generation Sunshine State Standards

Geometry	
Standard 1	Points, Lines, Planes, and Angles
Standard 2	Polygons
Standard 3	Quadrilaterals

(continued)

Standard 4	Triangles
Standard 5	Right Triangles
Standard 6	Polyhedrons and Other Solids
Standard 7	Mathematical Reasoning and Problem Solving
Standard 8	Logic
Standard 9	Circles
Strand B	Discrete Mathematics
Strand C	Geometry

Benchmarks

The following are the specific standards and benchmarks of the Geometry EOC Assessment.

Standard 1: Points, Lines, Planes and Angles

MA.912.G.1.1 (Moderate) Find the lengths and midpoints of line segments in two-dimensional coordinate systems.

MA.912.G.1.3 (Moderate) Identify and use the relationships between special pairs of angles formed by parallel lines and transversals.

Standard 2: Polygons

MA.912.G.2.1 (High) Identify and describe convex, concave, regular, and irregular polygons.

MA.912.G.2.2 (Moderate) Determine the measures of interior and exterior angles of polygons, justifying the method used.

MA.912.G.2.3 (High) Determine the measures of interior and exterior angles of polygons, justifying the method used.

MA.912.G.2.4 (High) Apply transformations (translations, reflections, rotations, dilations, and scale factors) to polygons to determine congruence, similarity, and symmetry. Know that images formed by translations, reflections, and rotations are congruent to the original shape. Create and verify tessellations of the plane using polygons.

MA.912.G.2.5 (Moderate) Explain the derivation and apply formulas for perimeter and area of polygons (triangles, quadrilaterals, pentagons, etc.).

MA.912.G.2.7 (Moderate) Determine how changes in dimensions affect the perimeter and area of common geometric figures.

Standard 3: Quadrilaterals

MA.912.G.3.1 (Moderate) Describe, classify, and compare relationships among quadrilaterals including the square, rectangle, rhombus, parallelogram, trapezoid, and kite.

MA.912.G.3.2 (Moderate) Compare and contrast special quadrilaterals on the basis of their properties.

MA.912.G.3.3 (High) Use coordinate geometry to prove properties of congruent, regular, and similar quadrilaterals.

MA.912.G.3.4 (High) Prove theorems involving quadrilaterals.

Standard 4: Triangles

MA.912.G.4.1 (Moderate) Classify, construct, and describe triangles that are right, acute, obtuse, scalene, isosceles, equilateral, and equiangular.

MA.912.G.4.2 (Moderate) Define, identify, and construct altitudes, medians, angle bisectors, perpendicular bisectors, orthocenter, centroid, incenter, and circumcenter.

MA.912.G.4.3 (High) Construct triangles congruent to given triangles.

MA.912.G.4.4 (Moderate) Use properties of congruent and similar triangles to solve problems involving lengths and areas.

MA.912.G.4.5 (Moderate) Apply theorems involving segments divided proportionally.

MA.912.G.4.6 (High) Prove that triangles are congruent or similar and use the concept of corresponding parts of congruent triangles.

MA.912.G.4.7 (Moderate) Apply the inequality theorems: triangle inequality, inequality in one triangle, and the Hinge Theorem.

Standard 5: Right Triangles

MA.912.G.5.1 (High) Prove and apply the Pythagorean Theorem and its converse.

MA.912.G.5.2 (Moderate) State and apply the relationships that exist when the altitude is drawn to the hypotenuse of a right triangle.

MA.912.G.5.3 (Moderate) Use special right triangles (30° - 60° - 90° and 45° - 45° - 90°) to solve problems.

MA.912.G.5.4 (High) Solve real-world problems involving right triangles.

MA.912.T.2.1 (Moderate) Define and use the trigonometric ratios (sine, cosine, tangent, cotangent, secant, cosecant) in terms of angles of right triangles.

Standard 6: Polyhedrons and Other Solids

MA.912.G.7.1 :(Moderate) Describe and make regular, non-regular, and oblique polyhedra, and sketch the net for a given polyhedron and vice versa.

MA.912.G.7.2 (Moderate) Describe the relationships between the faces, edges, and vertices of polyhedra.

MA.912.G.7.4 (Low) Identify chords, tangents, radii, and great circles of spheres

MA.912.G.7.5 (Moderate) Explain and use formulas for lateral area, surface area, and volume of solids.

MA.912.G.7.6 (Moderate) Identify and use properties of congruent and similar solids.

MA.912.G.7.7 (Moderate) Determine how changes in dimensions affect the surface area and volume of common geometric solids.

Standard 7: Mathematical Reasoning and Problem Solving

MA.912.G.8.1 (High) Analyze the structure of Euclidean geometry as an axiomatic system. Distinguish between undefined terms, definitions, postulates, and theorems.

MA.912.G.8.2 (Moderate) Use a variety of problem-solving strategies, such as drawing a diagram, making a chart, guess-and-check, solving a simpler problem, writing an equation, and working backwards.

MA.912.G.8.3 (Moderate) Determine whether a solution is reasonable in the context of the original situation.

MA.912.G.8.4 (High) Make conjectures with justifications about geometric ideas. Distinguish between information that supports a conjecture and the proof of a conjecture.

MA.912.G.8.5: (High) Write geometric proofs, including proofs by contradiction and proofs involving coordinate geometry. Use and compare a variety of ways to present deductive proofs, such as flow charts, paragraphs, two-column, and indirect proofs.

Standard 8: Logic

MA.912.D.6.2 (Moderate) Find the converse, inverse, and contrapositive of a statement

MA.912.D.6.3 (Moderate) Determine whether two propositions are logically equivalent.

MA.912.D.6.4 (Moderate) Use methods of direct and indirect proof and determine whether a short proof is logically valid.

Standard 9: Circles

MA.912.G.6.2 (Low) Define and identify: circumference, radius, diameter, arc, arc length, chord, secant, tangent and concentric circles.

MA.912.G.6.4 (Moderate) Determine and use measures of arcs and related angles (central, inscribed, and intersections of secants and tangents).

MA.912.G.6.5 (High) Solve real-world problems using measures of circumference, arc length, and areas of circles and sectors.

MA.912.G.6.6 (Moderate) Given the center and the radius, find the equation of a circle in the coordinate plane or given the equation of a circle in center-radius form, state the center and the radius of the circle.

MA.912.G.6.7 (Moderate) Given the equation of a circle in center-radius form or given the center and the radius of a circle, sketch the graph of the circle.

Strand B: Discrete Mathematics

MA.912.D.7.1 Perform set operations such as union and intersection, complement, and cross product.

MA.912.D.7.2 Use Venn diagrams to explore relationships and patterns and to make arguments about relationships between sets.

Strand C: Geometry

MA.912.G.1.4 Use coordinate geometry to find slopes, parallel lines, perpendicular lines, and equations of lines.

Chapter 1

Points, Lines, Planes, and Angles

Your Goals for Chapter 1

1. You should be able to apply the distance formula to find the length of line segments in coordinate systems.
2. You should be able to find the midpoints of line segments in coordinate systems.
3. You should be able to find angle measures formed by parallel lines and transversals.

Standards

The following standards are assessed on Florida's Geometry End-of-Course Assessment either directly or indirectly:

MA.912.G.1.1 (Moderate) Find the lengths and midpoints of line segments in two-dimensional coordinate systems.

MA.912.G.1.3 (Moderate) Identify and use the relationships between special pairs of angles formed by parallel lines and transversals.

These standards will be tested with a variety of questions that are characterized as being of low, moderate, and high difficulty. Difficulty is different from complexity. Don't be fooled: just because a problem has several steps does not mean that its difficulty level is high. All of the steps of a particular problem could be considered of "low" difficulty, which means that the problem is complex and time consuming, but not necessarily difficult. Difficulty is determined by how much thinking or background knowledge is involved in solving the problem.

So what do you actually need to know? You need to be able to apply the distance and midpoint formulas to solve problems. You also need to know the special pairs of angles formed by parallel lines and a transversal and use the relationships to find missing angles.

Distance Formula

Given the two points (x_1, y_1) *and* (x_2, y_2), the distance between these two points is given by the formula:

$$d = \sqrt{(x_2 - x_1)^2 + (y_2 - y_1)^2}$$

The subscripts indicate that there is a first and second point. It does not matter which point is first and which one is second, the answer will be the same.

Example:

Find the distance between the points (0, –3) and (–2, 4).

Using (0, –3) as (x_1, y_1) and (–2, 4) as (x_2, y_2), substitute into the formula.

$$d = \sqrt{(x_2 - x_1)^2 + (y_2 - y_1)^2}$$

$$d = \sqrt{(-2 - 0)^2 + (4 - (-3))^2}$$

$$d = \sqrt{(-2)^2 + (7)^2}$$

$$d = \sqrt{4 + 49}$$

$$d = \sqrt{53} \approx 7.28$$

The distance between the points (0, –3) and (–2, 4) is approximately 7.28.

Example:

Find the length of $\overline{AB}$.

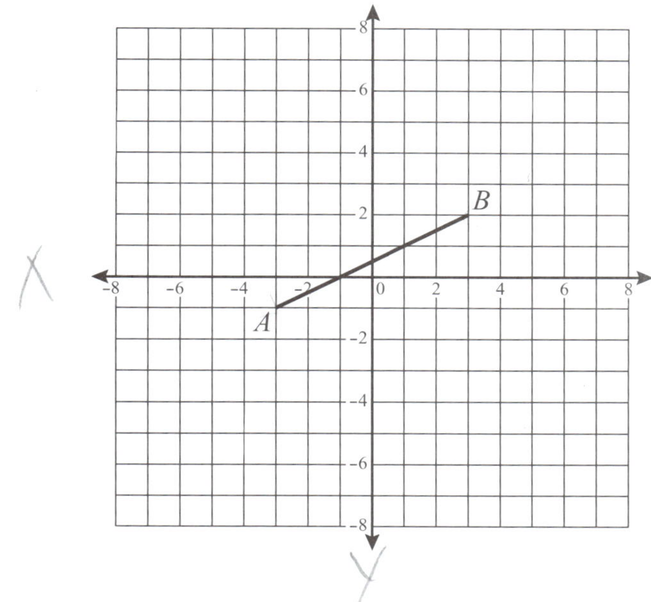

Identify the coordinates of point A and point B.

$A = (-3, -1)$ $B = (3, 2)$

Using $(-3, -1)$ as (x_1, y_1) and $(3, 2)$ as (x_2, y_2), substitute into the formula.

$$d = \sqrt{(x_2 - x_1)^2 + (y_2 - y_1)^2}$$

$$d = \sqrt{(3-(-3))^2 + (2-(-1))^2}$$

$$d = \sqrt{(6)^2 + (3)^2}$$

$$d = \sqrt{36 + 9}$$

$$d = \sqrt{45} \approx 6.71$$

The distance between the points $(-3, -1)$ and $(3, 2)$ is approximately 6.71.

Midpoint Formula

Given the two points (x_1, y_1) and (x_2, y_2), the midpoint between these two points is given by the formula:

$$M = \left(\frac{x_1 + x_2}{2}, \frac{y_1 + y_2}{2} \right)$$

The subscripts indicate that there is a first and second point. It does not matter which point is first and which one is second, the answer will be the same.

Example:

Find the midpoint of a segment with endpoints (–4, –6) and (0, –8).

Using (–4, –6) as (x_1, y_1) and (0, –8) as (x_2, y_2), substitute into the formula.

$$M = \left(\frac{x_1 + x_2}{2}, \frac{y_1 + y_2}{2} \right)$$

$$M = \left(\frac{-4 + 0}{2}, \frac{-6 + (-8)}{2} \right)$$

$$M = \left(\frac{-4}{2}, \frac{-14}{2} \right)$$

$$M = (-2, -7)$$

The midpoint of a segment with endpoints (–4, –6) and (0, –8) is (–2, –7).

Example:

Find the midpoint of $\overline{EF}$.

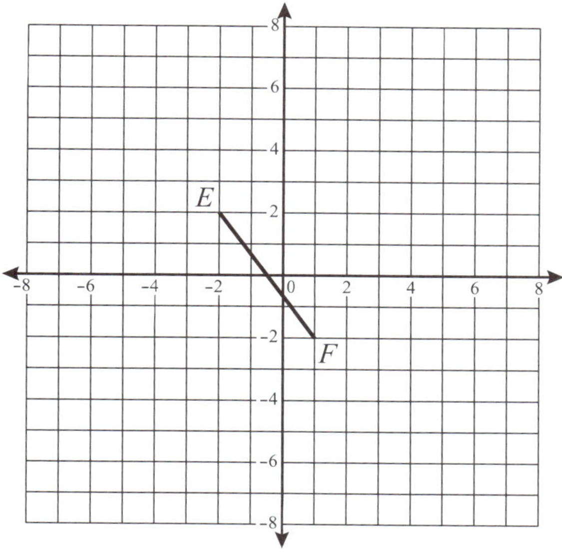

Identify the coordinates of point E and point F.

$E = (-2, 2)$ $F = (1, -2)$

Using (–2, 2) as (x_1, y_1) and (1, –2) as (x_2, y_2), substitute into the formula.

$$M = \left(\frac{x_1 + x_2}{2}, \frac{y_1 + y_2}{2}\right)$$

$$M = \left(\frac{-2 + 1}{2}, \frac{2 + (-2)}{2}\right)$$

$$M = \left(\frac{-1}{2}, \frac{0}{2}\right)$$

$$M = \left(-\frac{1}{2}, 0\right)$$

The midpoint of a segment with endpoints (–2, 2) and (1, –2) is $\left(-\frac{1}{2}, 0\right)$.

Example:

On a coordinate plane, $\overline{XY}$ has endpoint Y at (8, 4). The midpoint of $\overline{XY}$ is M (2, –1). What are the coordinates of point X?

Identify the coordinates of point X, point Y and the midpoint M.

$X = (x, y)$ $\quad$ $Y = (8, 4)$ $\quad$ $M = (2, -1)$

Using (x, y) as (x_1, y_1), (8, 4) as (x_2, y_2) and (2, –1) as M, substitute into the formula:

$$M = \left(\frac{x_1 + x_2}{2}, \frac{y_1 + y_2}{2}\right)$$

$$(2, -1) = \left(\frac{x + 8}{2}, \frac{y + 4}{2}\right)$$

Now, separate the formula into two equations: one equation for the x-coordinate and one equation for the y-coordinate.

The equation to find the x-coordinate is:

$$2 = \frac{x + 8}{2}$$

$$2 \cdot 2 = \left(\frac{x + 8}{2}\right) \cdot 2$$ Multiply both sides by 2 to eliminate the denominator

$$4 = x + 8$$

$$4 - 8 = x + 8 - 8$$ Subtract 8 from both sides

$$-4 = x$$

The equation to find the y-coordinate is:

$-1 = \frac{y+4}{2}$

$-1 \cdot 2 = \left(\frac{y+4}{2}\right) \cdot 2$ Multiply both sides by 2 to eliminate the denominator

$-2 = y + 4$

$-2 - 4 = y + 4 - 4$ Subtract 4 from both sides

$-6 = y$

The coordinate of point X is $(-4, -6)$.

Angles Formed by a Transversal

A transversal is a line that intersects two lines in the same plane at different points.

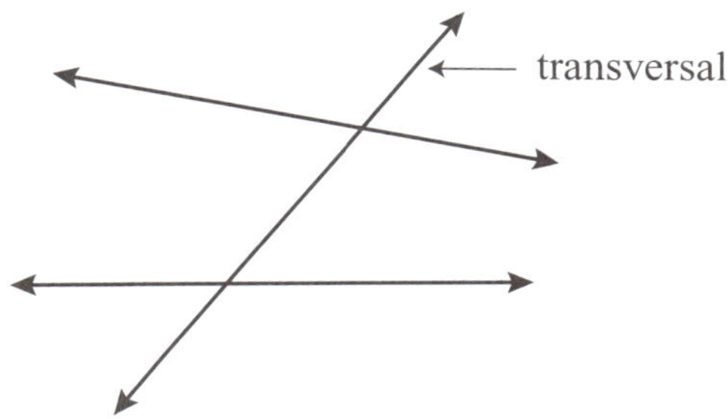

If the intersected lines are parallel, the transversal produces several special pairs of angles.

Congruent Angle Pairs	Supplementary Angle Pairs
Corresponding Angles Alternate Interior Angles Alternate Exterior Angles	Same Side Interior Angles

Line m is parallel to line n.

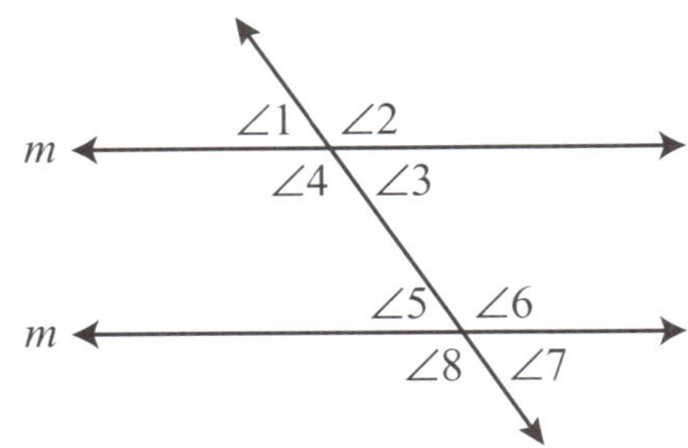

Corresponding Angles	**Alternate Interior Angles**
$\angle 1$ and $\angle 5$ $\angle 2$ and $\angle 6$ $\angle 3$ and $\angle 7$ $\angle 4$ and $\angle 8$	$\angle 3$ and $\angle 5$ $\angle 4$ and $\angle 6$
Alternate Exterior Angles	**Same Side Interior Angles**
$\angle 1$ and $\angle 7$ $\angle 2$ and $\angle 8$	$\angle 4$ and $\angle 5$ $\angle 3$ and $\angle 6$

Example:

Given line a is parallel to line b, determine the value of x.

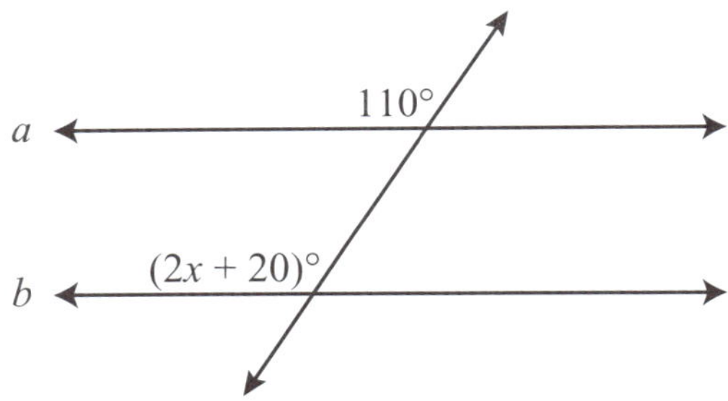

Line a is parallel to line b and the angles given are corresponding, therefore they are congruent. Since the angles are congruent, the following equation can be used to solve for x:

$2x + 20 = 110$

$2x + 20 - 20 = 110 - 20$ Subtract 20 from both sides

$2x = 90$

$\frac{2x}{2} = \frac{90}{2}$ Divide both sides by 2

$x = 45$

Example:

Given line a is parallel to line b, determine the measure of each angle.

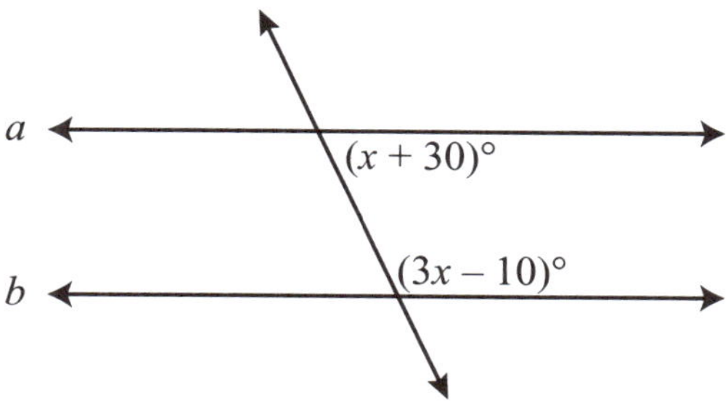

Line a is parallel to line b and the angles given are same side interior, therefore they are supplementary. Since the angles are supplementary, the following equation can be used to solve for x:

$(x + 30) + (3x - 10) = 180$

$4x + 20 = 180$ Add and subract like terms.

$4x + 20 - 20 = 180 - 20$ Subtract 20 from both sides.

$4x = 160$

$\frac{4x}{4} = \frac{160}{4}$ Divide both sides by 4

$x = 40$

Substitute $x = 40$ into each expression.

$(x + 30)°$	$(3x - 10)°$
$(40 + 30)°$	$(3(40) - 10)°$
$70°$	$(120 - 10)°$
	$110°$

The angles in the figure measure 70° and 110°.

End-of-Chapter Quiz

1. Use the distance formula to determine the relationship between the length of $\overline{XY}$ and the length of $\overline{YZ}$ given X (0, –8), Y (4, 3), Z (–2, –7).

A. $\overline{XY} = \overline{YZ}$

B. $\overline{XY} < \overline{YZ}$

C. $\overline{XY} > \overline{YZ}$

D. No relationship can be determined

2. Find the midpoint of $\overline{XY}$ with endpoints X (–4, 0) and Y (3, 5).

A. $\left(\frac{1}{2}, \frac{5}{2}\right)$

B. $\left(\frac{7}{2}, \frac{5}{2}\right)$

C. $\left(\frac{-7}{2}, \frac{5}{2}\right)$

D. $\left(\frac{-1}{2}, \frac{5}{2}\right)$

3. On a coordinate plane, $\overline{XY}$ has endpoint Y at (4, 3). The midpoint of $\overline{XY}$ is M (–1, –2). What are the coordinates of point X?

A. (–6, –7)

B. (–6, –1)

C. (2, –7)

D. (2, –1)

4. **Find the perimeter of the triangle given on the coordinate plane.**

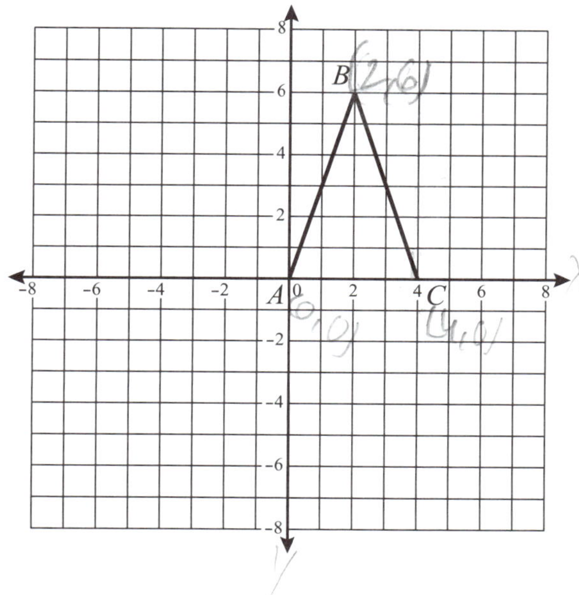

A. $4 + 4\sqrt{2}$

B. $4 + 4\sqrt{10}$

C. $5 + 4\sqrt{2}$

D. $5 + 4\sqrt{10}$

5. **Find the length of $\overline{CD}$ given C (3, 2) and D (–1, 4).**

A. $2\sqrt{2}$

B. $2\sqrt{5}$

C. $2\sqrt{10}$

D. $2\sqrt{13}$

6. **If X is the midpoint of $\overline{AB}$ and $AX = 12$, find the length of $\overline{AB}$.**

A. 6

B. 12

C. 18

D. 24

7. **$\overline{AB}$ is a diameter of the circle. Determine the coordinates of the center of the circle.**

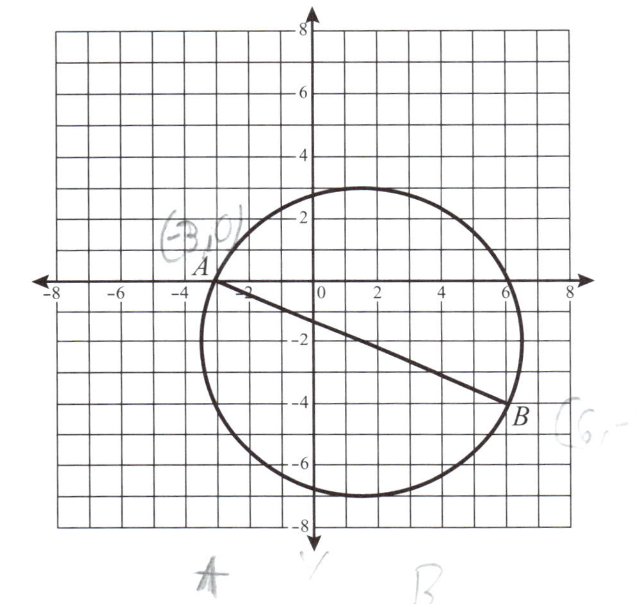

A. (–4.5, –2)

B. (–4.5, 2)

C. (1.5, –2)

D. (1.5, 2)

8. If $m\angle A = (2x - 4)$ and $m\angle B = 100°$, find the value of x. Assume that line m is parallel to line n.

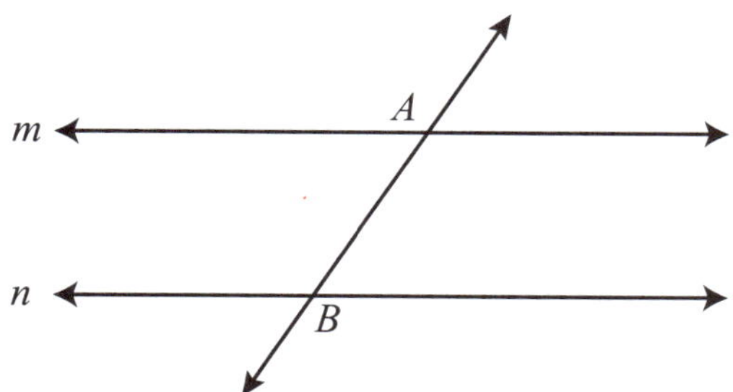

A. $x = 48$

B. $x = 52$

C. $x = 96$

D. $x = 104$

9. If line a is parallel to line b, which of the following statements is true?

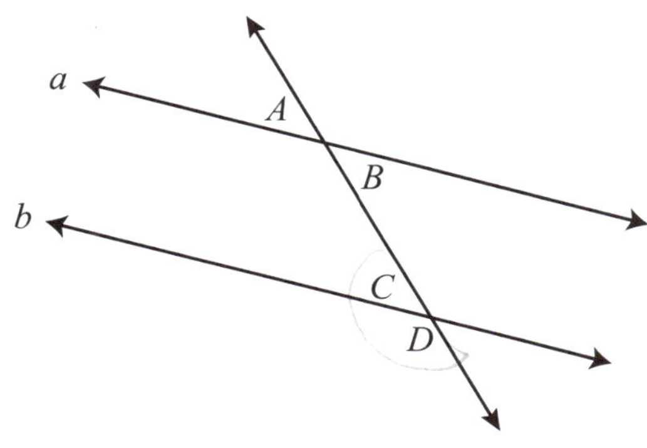

A. $m\angle A + m\angle C = 180°$

B. $m\angle B + m\angle C = 180°$

C. $\angle A \cong \angle D$

D. $\angle B \cong \angle C$

10. In the diagram, line *m* and line *n* are parallel. Find $m\angle ABC$.

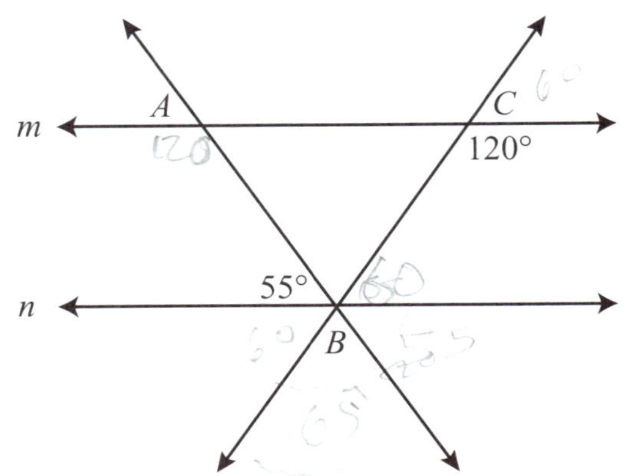

A. 55°

B. 65°

C. 120°

D. 175°

Answers to the Chapter Quiz

1. C

$XY = \sqrt{(0-4)^2 + (-8-3)^2}$

$XY = \sqrt{16+121}$

$XY = \sqrt{137}$

$YZ = \sqrt{(4+2)^2 + (3+7)^2}$

$YZ = \sqrt{36+100}$

$YZ = \sqrt{136}$

$\sqrt{137} > \sqrt{136}$, therefore $XY > YZ$

2. D

$$M = \left(\frac{-4+3}{2}, \frac{0+5}{2} \right)$$

$$M = \left(\frac{-1}{2}, \frac{5}{2} \right)$$

3. A

$$(-1,-2) = \left(\frac{x+4}{2}, \frac{y+3}{2} \right)$$

$$-1 = \frac{x+4}{2} \qquad -2 = \frac{y+3}{2}$$

$$-2 = x+4 \qquad -4 = y+3$$

$$-6 = x \qquad -7 = y$$

$(-6, -7)$

4. B

To find the perimeter, the length of each side needs to be added together.

By counting the spaces, $AC = 4$.

Since the triangle is isosceles, $AB \cong BC$. Use the distance formula to find the length of one of these sides.

$$AB = \sqrt{(2-0)^2 + (6-0)^2}$$

$$AB = \sqrt{4+36}$$

$$AB = \sqrt{40} = 2\sqrt{10}$$

$AC = 4$, $AB = 2\sqrt{10}$ and $BC = 2\sqrt{10}$, so the perimeter of the triangle is $4 + 2\sqrt{10} + 2\sqrt{10} = 4 + 4\sqrt{10}$

5. B

$\sqrt{(3-(-1))^2+(2-4)^2}$

$\sqrt{16+4}$

$\sqrt{20}=2\sqrt{5}$

6. D

Since $AX = 12$, then $XB = 12$ because X is the midpoint of $\overline{AB}$. So, $AB = AX + XB = 12 + 12 = 24$.

7. C

The center of the circle is the midpoint of the diameter.

$\left(\frac{-3+6}{2}, \frac{0+(-4)}{2}\right)$

$(1.5, -2)$

8. B

$2x - 4 = 100$

$2x = 104$

$x = 52$

9. D

$\angle B$ and $\angle C$ are alternate interior angles, so the angles are congruent.

10. B

$55° + m\angle ABC = 120°$

$m\angle ABC = 65$

Chapter 2

Polygons

Your Goals for Chapter 2

1. You should be able to determine the measures on interior and exterior angles of polygons.

2. You should be able to use the properties of similar and congruent polygons.

3. You should be able to apply transformations.

4. You should be able to determine the perimeter and area of polygons and understand how a change in dimensions affects these measures.

Standards

The following standards are assessed on Florida's Geometry End-of-Course exam either directly or indirectly:

MA.912.G.2.1: (High) Identify and describe convex, concave, regular, and irregular polygons.

MA.912.G.2.2: (Moderate) Determine the measures of interior and exterior angles of polygons, justifying the method used.

MA.912.G.2.3: (High) Use properties of congruent and similar polygons to solve mathematical or real-world problems.

MA.912.G.2.4: (High) Apply transformations (translations, reflections, rotations, dilations, and scale factors) to polygons to determine congruence, similarity, and symmetry. Know that images formed by translations, reflections, and rotations are congruent to the original shape. Create and verify tessellations of the plane using polygons.

MA.912.G.2.5: (Moderate) Explain the derivation and apply formulas for perimeter and area of polygons (triangles, quadrilaterals, pentagons, etc.).

MA.912.G.2.7: (Moderate) Determine how changes in dimensions affect the perimeter and area of common geometric figures.

Convex vs. Concave

A **polygon** is a two-dimensional closed shape with straight lines and is either convex or concave. **Convex polygons** are polygons whose interior angles are each less than 180°. That is, no angles point inward. Any polygon that is not convex is considered a **concave polygon**.

Regular vs. Irregular

A **regular polygon** is a polygon with all equal angles and sides. If a polygon does not have equal angles and sides, then it is an **irregular polygon**.

Interior Angles of Polygons

The **sum of the interior angles** of a polygon can be found using the following formula:

$(n - 2)180°$

where n is the number of sides in the polygon.

number of sides	sum of the interior angles
3	$(3 - 2)180° = 180°$
4	$(4 - 2)\ 180° = 360°$
5	$(5 - 2)180° = 540°$
⋮	
n	$(n - 2)180°$

Example:

Find the sum of the interior angles of a 12-sided polygon.

$(n - 2)180°$

$(12 - 2)180°$

$(10)180°$

$1800°$

The sum of the interior angles of a 12-sided figure is 1800°.

Example:

Find the measure of each unknown angle, x.

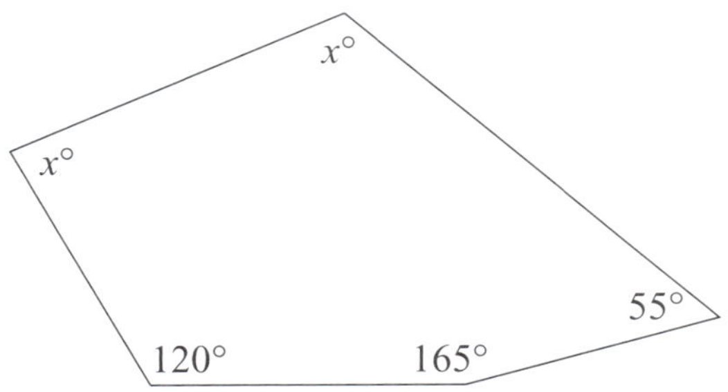

First, find the sum of the interior angles of the polygon. Since the polygon has 5 sides, it is a pentagon. The sum of the interior angles is determined as follows:

$(n - 2)180°$

$(5 - 2)180°$

$(3)180°$

$540°$

The sum of the interior angles of the figure is 540°.

Using this sum, the following equation can be used to find the value of x:

$x° + x° + 120° + 165° + 55° = 540°$

$2x° + 340° = 540°$

$2x° + 340° - 340° = 540° - 340°$

$2x = 200°$

$\frac{2x°}{2} = \frac{200°}{2}$

$x = 100°$

Exterior Angles of Polygons

Regardless of the number of sides in the polygon, the **sum of the exterior angles** is always **360°**. An interior angle and one of the exterior angles at the same vertex are supplementary.

Example:

Steven is drawing a regular octagon. What is the measure of one exterior angle of the octagon?

The sum of the exterior angles of any polygon is 360°. Since Steven is drawing an octagon, the figure will have 8 exterior angles. Each exterior angle is congruent because the octagon is regular. To determine the measure of one of the exterior angles, the total sum, 360°, is divided by 8.

$$360° \div 8 = 45°$$

Each exterior angle of the regular octagon measures 45°.

EXERCISE 1

1. **Find the sum of the interior angles of a 16-sided polygon.**

2. Find the $m\angle 1$ in the figure below.

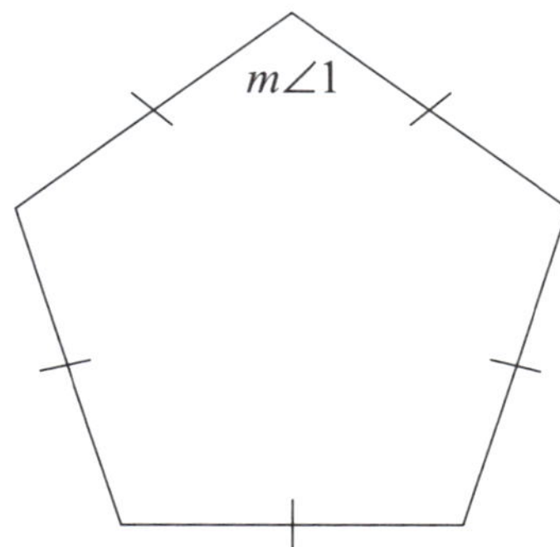

3. If the sum of the interior angles of a polygon is 2340°, how many sides does the polygon have?

4. If each exterior angle measures 20°, how many sides does the polygon have?

5. Find the $m\angle 1$ in the figure below.

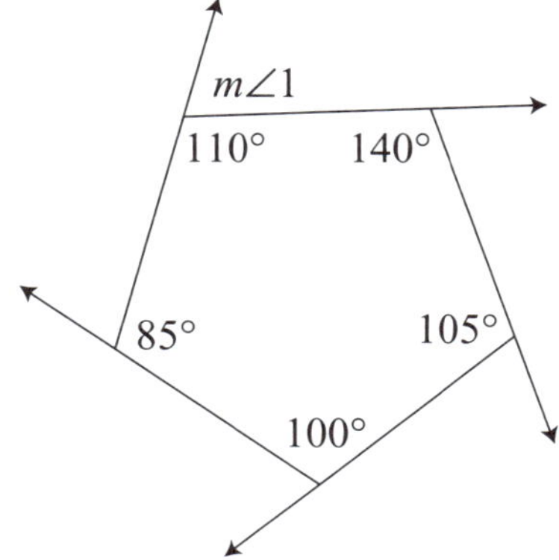

Congruent Polygons

Polygons that are the same shape and size are **congruent polygons**. Two polygons are congruent if all corresponding sides and angles are congruent.

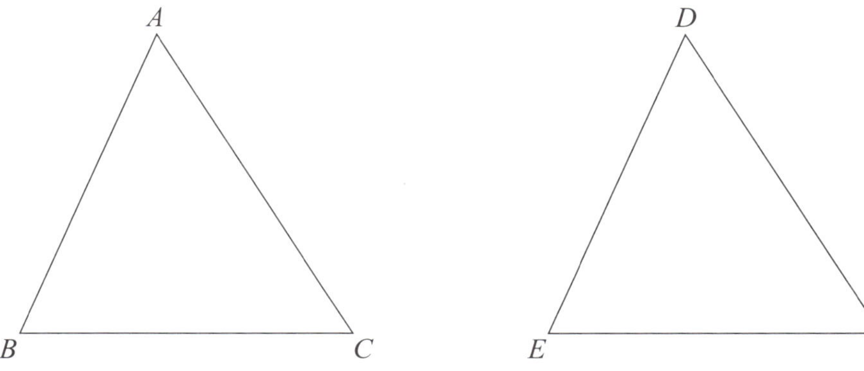

Corresponding Sides	Corresponding Angles
$\overline{AB}$ and $\overline{DE}$	$\angle A$ and $\angle D$
$\overline{BC}$ and $\overline{EF}$	$\angle B$ and $\angle E$
$\overline{AC}$ and $\overline{DF}$	$\angle C$ and $\angle F$

Example:

The polygons below *HKJI* and *LONM* are congruent. Name all pairs of congruent sides and angles.

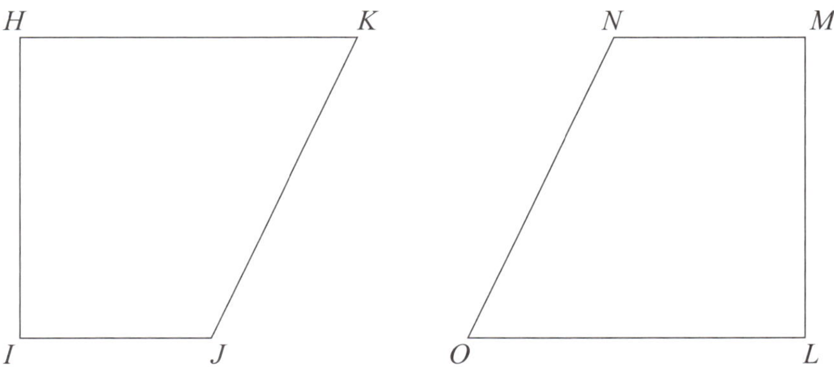

Since the figures are congruent, all corresponding sides and angles are congruent.

$HI \cong LM$	$\angle H \cong \angle L$
$IJ \cong MN$	$\angle I \cong \angle M$
$JK \cong NO$	$\angle J \cong \angle N$
$KH \cong OL$	$\angle K \cong \angle O$

Similar Polygons

Polygons that are the same shape, but a different size are **similar polygons**. Two polygons are similar if corresponding angles are congruent and corresponding sides have a proportional relationship. To have a proportional relationship, corresponding sides must change by the same scale factor. The ratio of two corresponding lengths in similar figures is the **scale factor**.

Example:

Given that the figures below are similar, determine the scale factor.

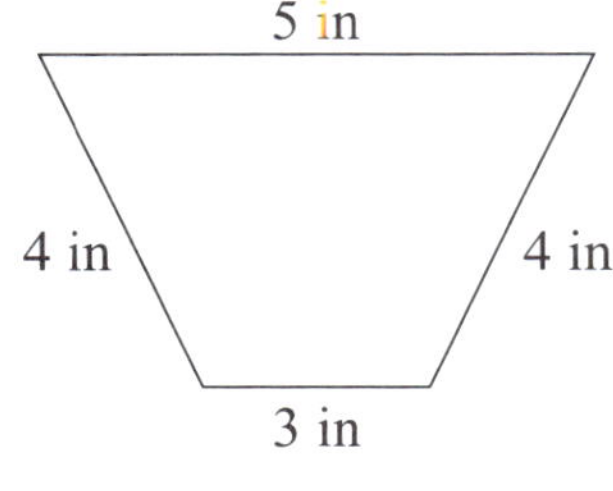

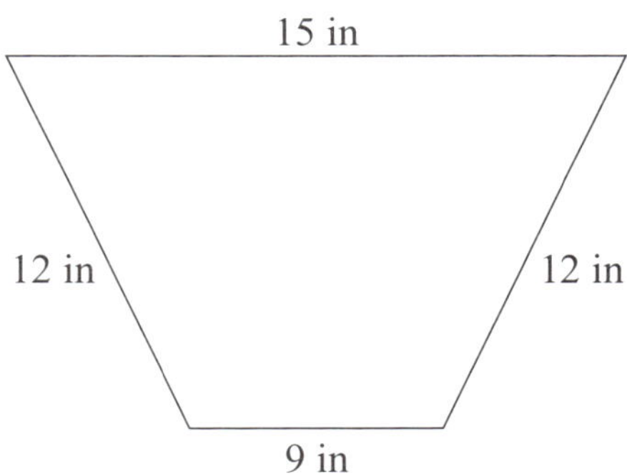

The scale factor is the ratio of two corresponding side lengths:

$$\frac{3}{9} = \frac{1}{3} \qquad \frac{4}{12} = \frac{1}{3} \qquad \frac{5}{13} = \frac{1}{3}$$

All pairs of corresponding sides are in ratio 1:3, therefore the scale factor is $\frac{1}{3}$.

Example:

Lori hired an architect to make a scale drawing of her home, but he forgot to label one of the lengths. Use the information given to find the missing length.

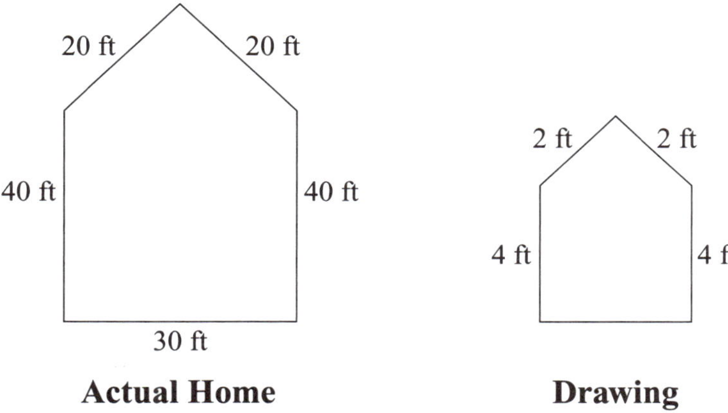

Scale drawings are similar figures to the actual object they represent. Therefore, the corresponding sides of the figures are proportional.

$$\frac{40 \text{ ft}}{4 \text{ ft}} = \frac{30 \text{ ft}}{x}$$

$$40x = 120 \quad \text{Cross multiply}$$

$$\frac{40x}{40} = \frac{120}{40} \quad \text{Divide both sides by 40}$$

$$x = 3 \text{ ft}$$

Therefore, the missing label in the drawing is 3 ft.

Transformations

A transformation changes the position of a figure. There are four types of transformations:

- rotation—turn
- reflection—flip
- translation—slide
- dilation—enlarge or reduce a figure

Rotations, reflections and translations result in images congruent to the original figure. Dilations result in images similar to the original figure.

Example:

Name the translation that transformed Rectangle *ABCD* to Rectangle *A′B′C′D′*.

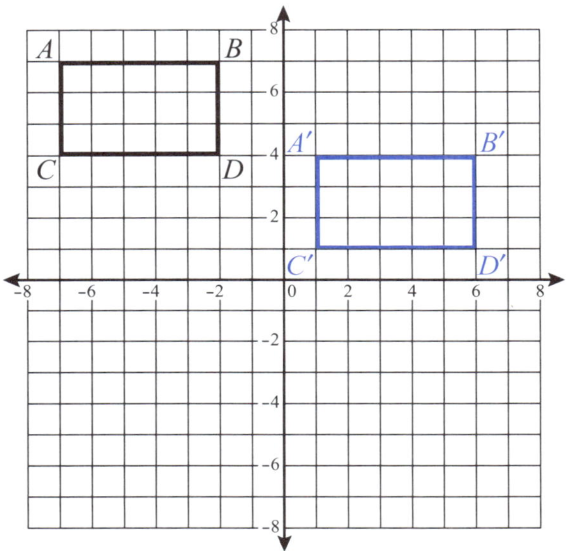

Rectangle *ABCD* has moved 8 units to the right and 3 units down. In coordinate notation:

$$(x, y) \rightarrow (x + 8, y - 3)$$

Remember that movement to the right is positive and left is negative. Movement up is positive and down is negative.

Example:

$\triangle XYZ$ is shown on the coordinate plane below. Reflect $\triangle XYZ$ over the line $x = 2$.

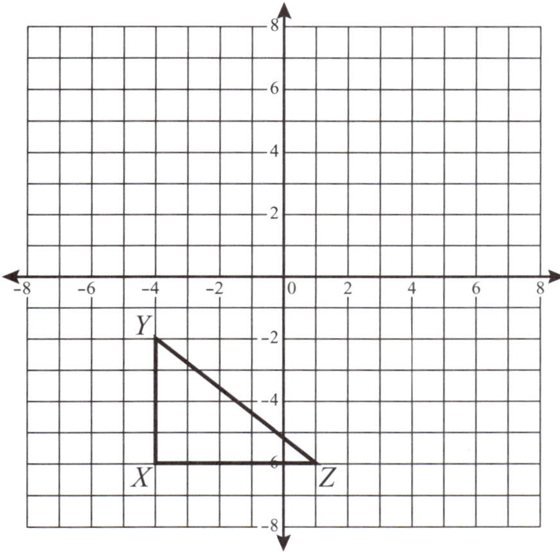

First identify the line $x = 2$. It is a vertical line passing through the (2, 0). Then flip the figure over that line.

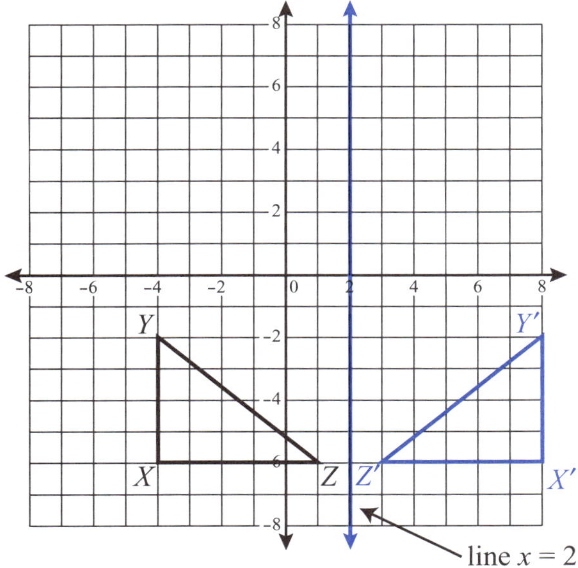

Example:

If Parallelogram *EFGH* is rotated 90° counterclockwise, determine the coordinate of H'.

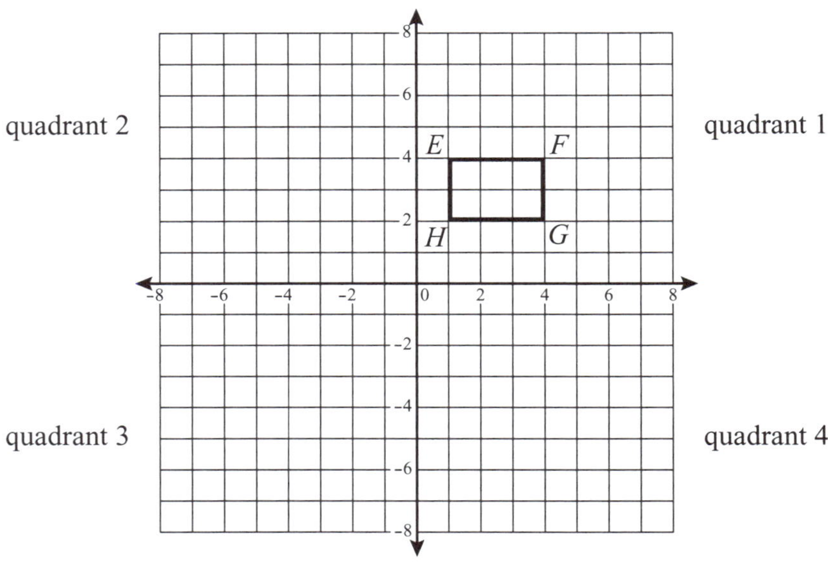

Parallelogram *EFGH* is in quadrant 1 on the coordinate plane. When the rectangle is rotated 90° counterclockwise, it will move to quadrant 2.

In general, when a figure is rotated counterclockwise about the origin, the following rules hold:

90° counterclockwise rotation: $(x, y) \rightarrow (-y, x)$

180° counterclockwise rotation: $(x, y) \rightarrow (-x, -y)$

270° counterclockwise rotation: $(x, y) \rightarrow (y, -x)$

Therefore, the image of $H(1, 2)$ after a 90° counterclockwise rotation is $H'(-2, 1)$.

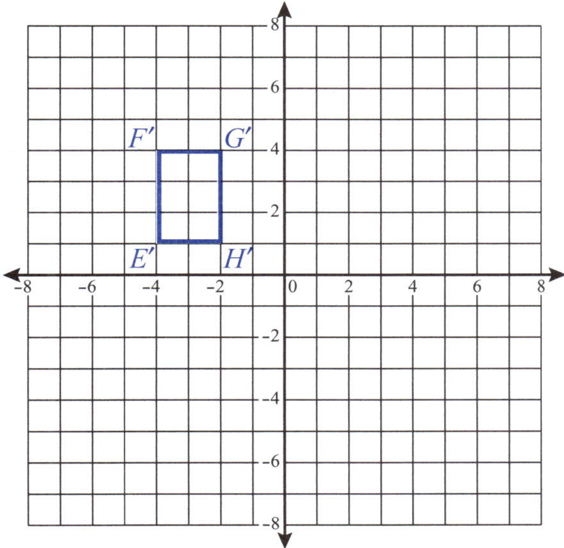

Example:

Dilate figure $LMNO$, with the origin as the center, using a scale factor of 2.

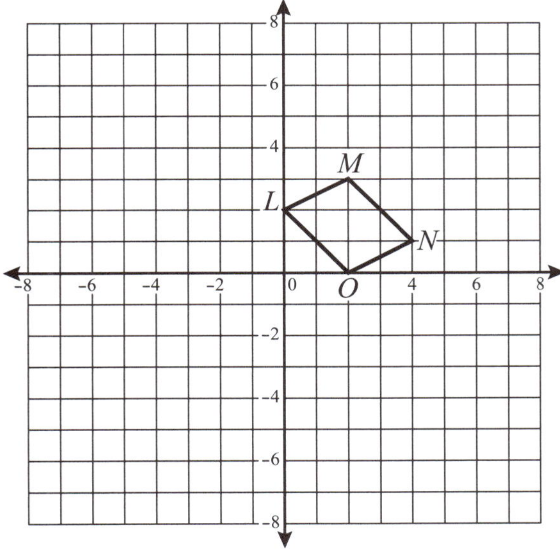

When dilating a figure with the origin as the center, each coordinate is multiplied by the scale factor.

$(x, y) \rightarrow (kx, ky)$ where k is the scale factor

Therefore, the coordinates of the image are:

L': $(0, 2) \rightarrow (2(0), 2(2)) \rightarrow (0, 4)$

M': $(2, 3) \rightarrow (2(2), 2(3)) \rightarrow (4, 6)$

N': $(4, 1) \rightarrow (2(4), 2(1)) \rightarrow (8, 2)$

O': $(2, 0) \rightarrow (2(2), 2(0)) \rightarrow (4, 0)$

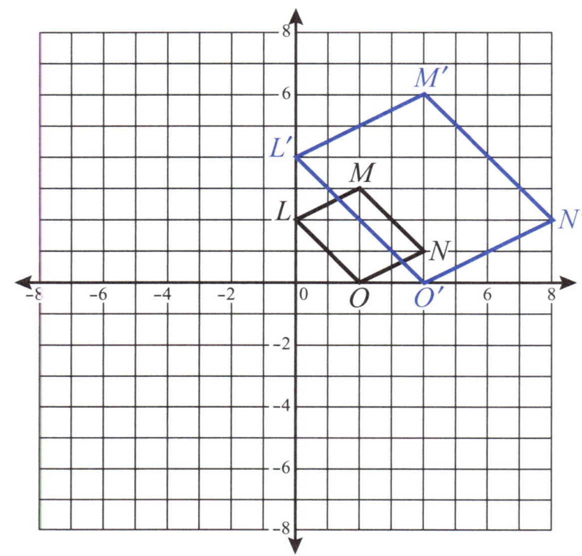

If dilation does not occur with center at the origin, translate the center of dilation and the figure so the origin becomes the center, and then translate it back after the dilation.

Exercise 2

1. Given Trapezoid *ABCD* ≅ Trapezoid *WXYZ*, which side in Trapezoid *WXYZ* is congruent to $\overline{AB}$?

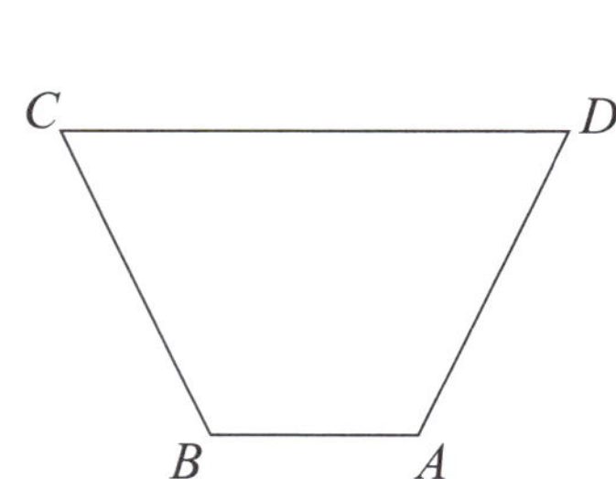

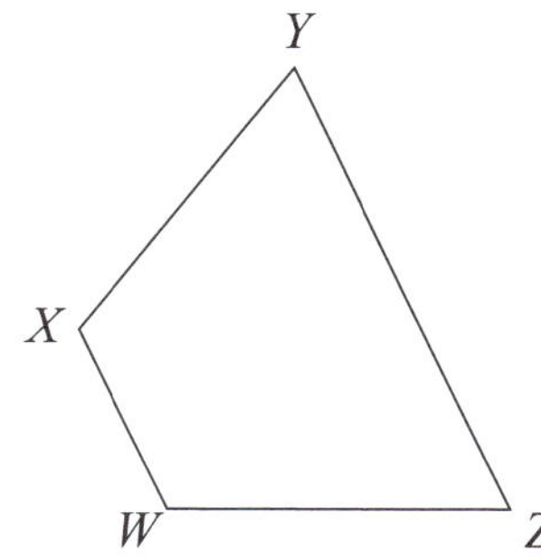

2. Given that the figures below are similar, find $m\angle 1$. $ABCDE \sim PQRST$.

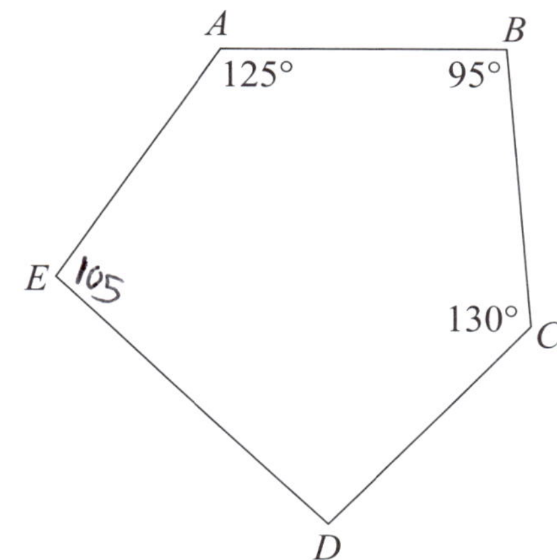

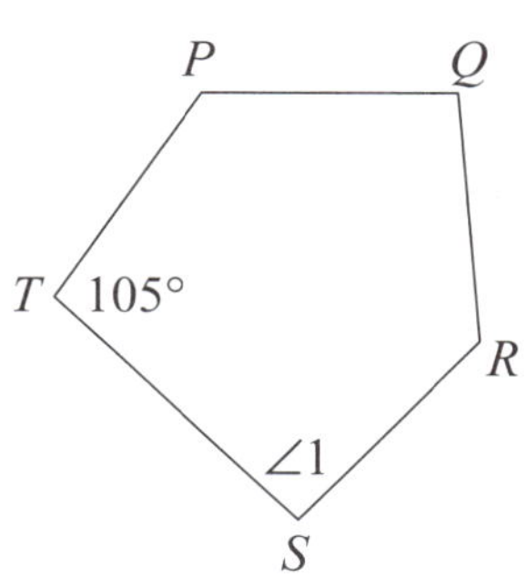

3. If $\triangle ABC$ is translated as follows:

$(x, y) \rightarrow (x - 2, y + 3)$,

what are the coordinates of A'?

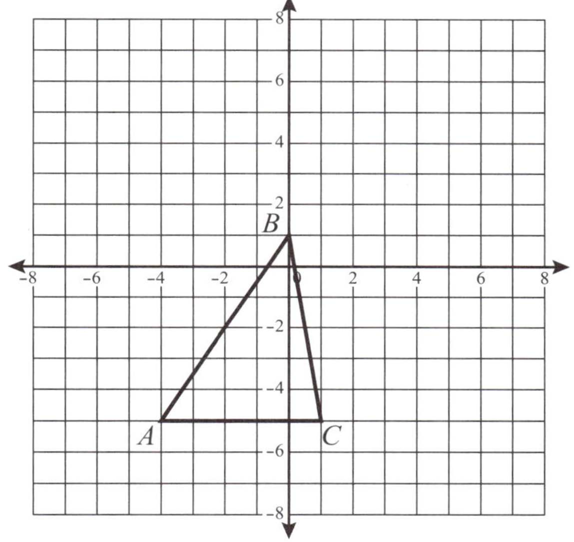

4. Reflect Parallelogram *EFGH* over the line $y = -1$.

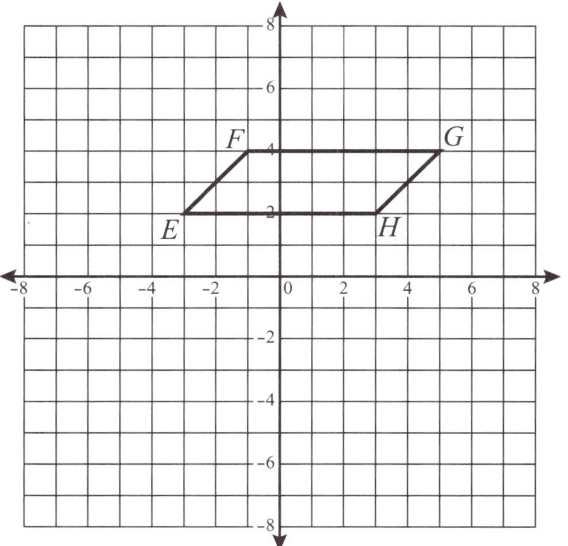

5. Determine the scale factor of the dilation. The image is the larger rectangle.

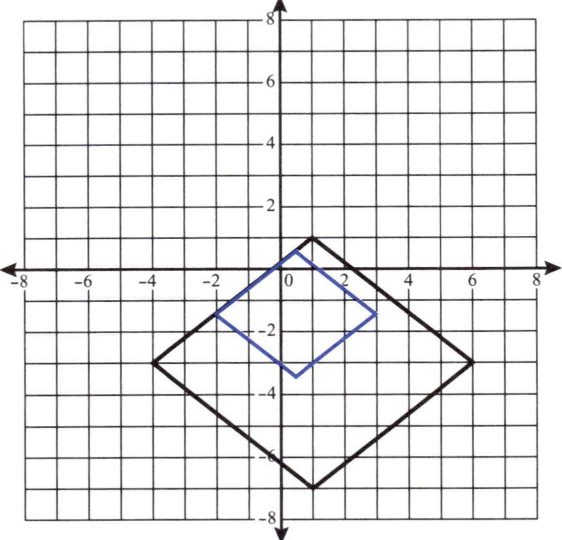

Perimeter and Area

Perimeter is the length around the outside of a figure. Perimeter is found by adding the length of each external side of a figure.

	Area Formula
Square	$A = s^2$; s = side length
Rectangle	$A = bh$; b = base, h = height
Triangle	$A = \frac{1}{2}bh$
Parallelogram	$A = bh$; b = base, h = height
Rhombus	$A = \frac{1}{2}d_1d_2$; d = length of diagonal
Trapezoid	$A = \frac{1}{2}h\ (b_1 + b_2)$; b = base, h = height
Regular Polygon	$A = \frac{1}{2}$(apothem) (perimeter)

Area is the amount of space in a plane inside a figure. Most common figures have a formula to find their area.

Note: Apothem is the perpendicular distance from the center of a regular polygon to a side. Height is the perpendicular distance from a vertex to the opposite side.

Example:

Find the perimeter of the figure.

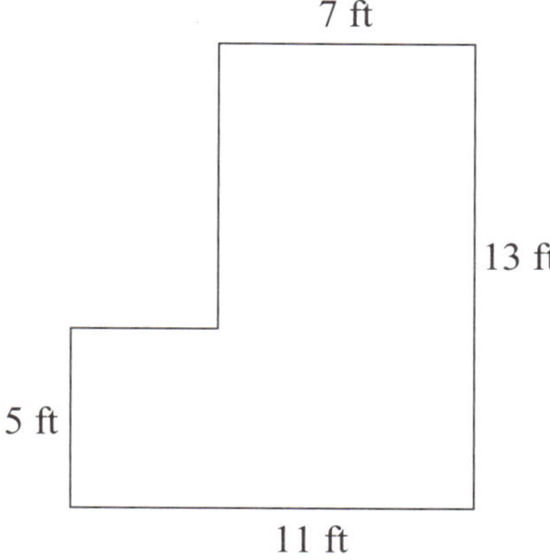

Perimeter is the length around the outside of a figure. To determine the perimeter of the above figure, the length of each side must be added together. The figure has 6 sides, but only 4 are labeled.

The unknown horizontal side can be found by subtracting 7 ft from the length of the entire parallel side, which is 11 ft. Thus, the unknown horizontal length is 4 ft.

The unknown vertical side can be found by subtracting 5 ft from the length of the entire parallel side, which is 13 ft. Thus, the unknown vertical length is 8 ft.

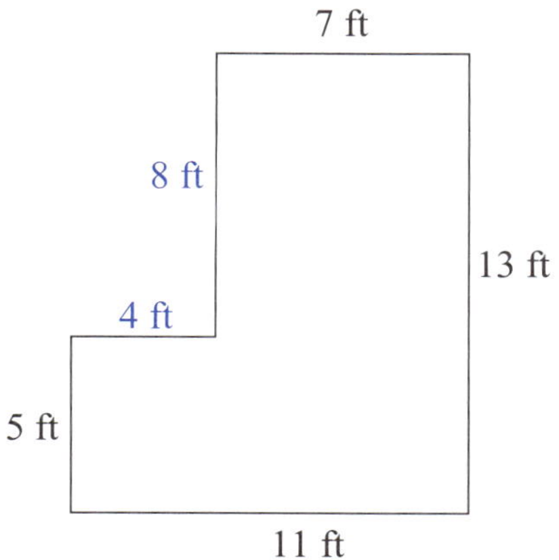

Perimeter = 5 ft + 11 ft + 13 ft + 7 ft + 8 ft + 4 f 8 ft

Example:

Find the area of the figure.

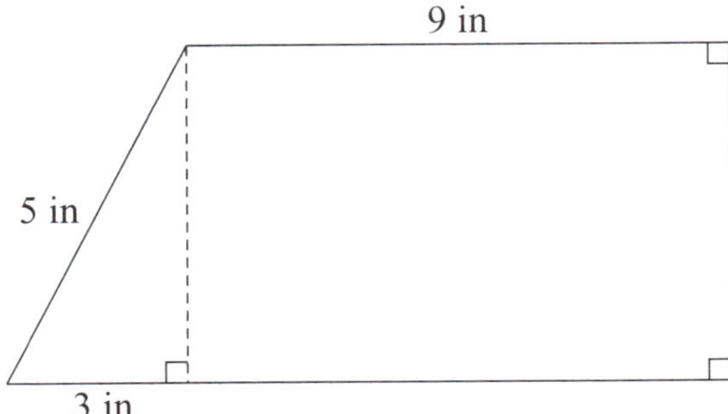

The figure given is a trapezoid. The formula for area of a trapezoid is:

$$A = \frac{1}{2}h(b_1 + b_2)$$

The height of the trapezoid is not given, but can be found using the Pythagorean Theorem with the right triangle formed by the dotted line drawn in for the height.

$$a^2 + b^2 = c^2$$

$$a^2 + 3^2 = 5^2$$

$$a^2 + 9 = 25$$

$$a^2 + 9 - 9 = 25 - 9$$

$$a^2 = 16$$

$$\sqrt{a^2} = \sqrt{16}$$

$$a = 4$$

The height of the trapezoid is 4 in.

The length of the bottom base is found by adding 3 in. and 9 in., which is 12 in.

Now substituting into the formula:

$$A = \frac{1}{2} \cdot h(b_1 + b_2)$$

$$A = \frac{1}{2} \cdot 4(9 + 12)$$

$$A = \frac{1}{2} \cdot 4(21)$$

$$A = 42 \text{ in}^2$$

Example:

Find the area of the regular pentagon.

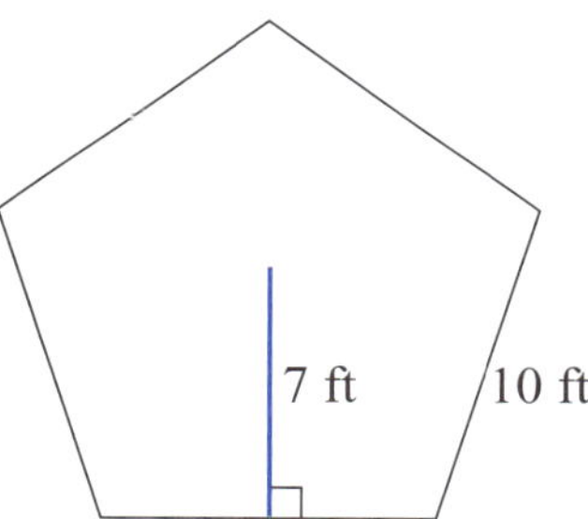

The line segment given in the figure is the apothem. The **apothem** is a line segment drawn from the center of a regular polygon that is perpendicular to one of its sides. The point where the apothem intersects is the midpoint of the side.

A regular pentagon is made up of 5 congruent triangles.

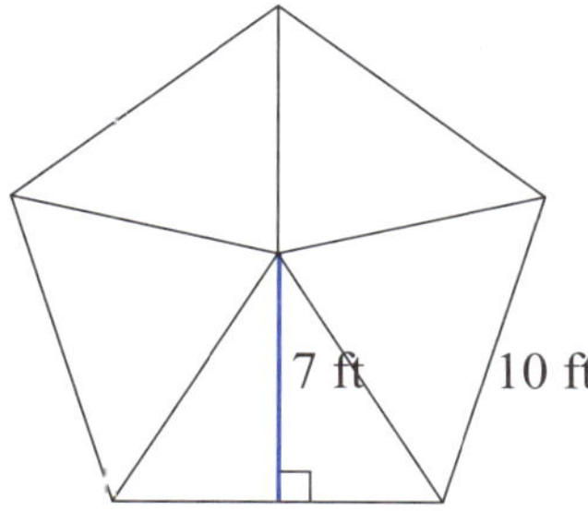

The formula for area of a triangle is:

$$A = \frac{1}{2}bh$$

$$A = \frac{1}{2}(10)(7)$$

$$A = \frac{1}{2}(70)$$

$$A = 35 \text{ ft}^2$$

The area of one triangle is then multiplied by 5 to account for each triangle.

$$A = 35(5) = 175 \text{ ft}^2$$

The area of a regular polygon can also be found using the formula:

$$A = \frac{1}{2}(\text{apothem})\ (\text{perimeter})$$

$$A = \frac{1}{2}(7)(50)$$

$$A = \frac{1}{2}(350)$$

$$A = 175 \text{ ft}^2$$

Perimeter and Area of Similar Polygons

If two polygons are similar with corresponding side lengths in the ratio $a : b$, then:

- the ratio of their perimeter is $a : b$, and
- the ratio of their areas is $a^2 : b^2$.

Example:

The parallelograms below are similar. Find the ratio of their areas.

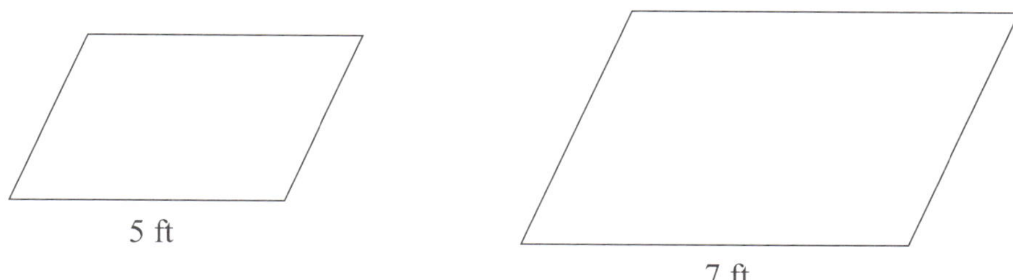

The ratio of the corresponding sides lengths is 5 : 7. Since the figures are similar, the ratio of the areas is $5^2 : 7^2$ or 25 : 49.

Example:

Rectangles *LMNO* and *PQRS* are similar. The area of *LMNO* is 12 in^2 and the area of *PQRS* is 27 in^2. What is the ratio of their perimeters?

The ratio of the areas is 12 : 27, which simplifies to 4 : 9. Therefore,

$$a^2 = 4 \qquad b^2 = 9$$

$$\sqrt{a^2} = \sqrt{4} \qquad \sqrt{b^2} = \sqrt{9}$$

$$a = 2 \qquad b = 3$$

The ratio of the perimeters is 2 : 3.

Exercise 3

1. **Find the perimeter of the parallelogram.**

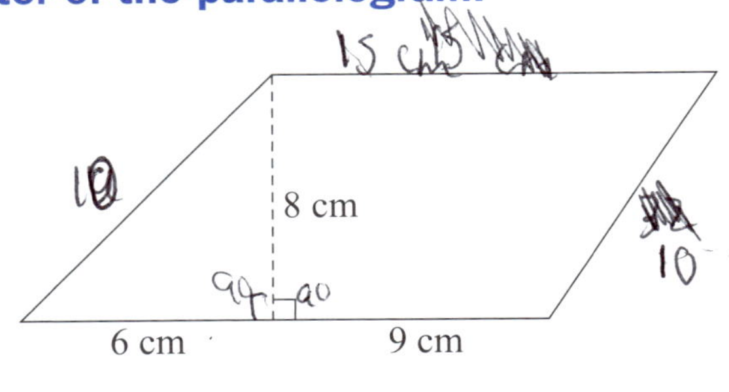

2. **Find the area of the regular hexagon.**

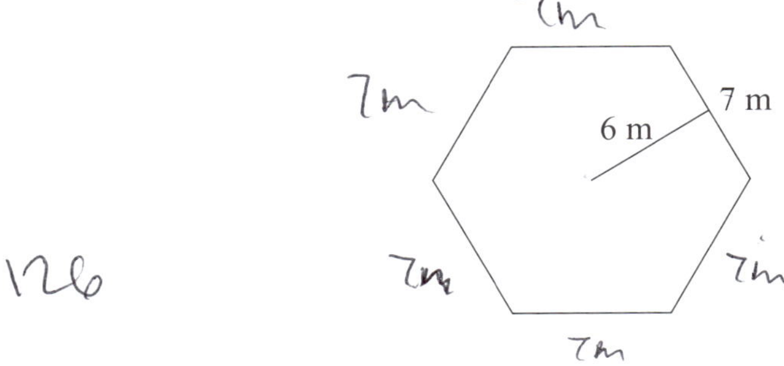

3. **The area of the rhombus is 60 ft^2. If the length of one diagonal is 10 ft, what is the length of the other diagonal?**

4. **The figures below are similar. What is the ratio of their areas?**

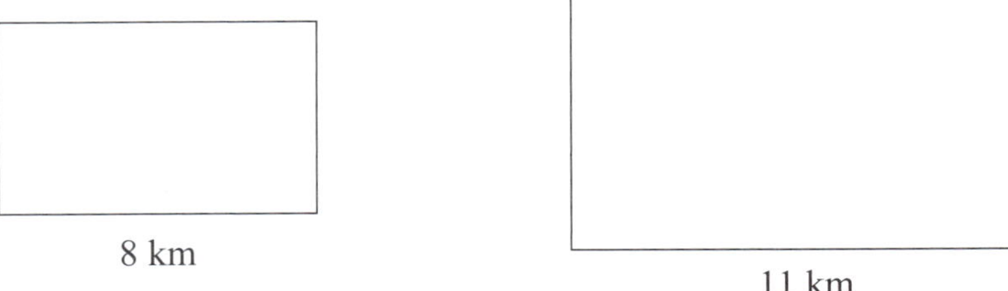

5. When Beth walked into class, there were two similar rectangles on the board. The area of rectangle *A* was 100 in^2 and the area of rectangle *B* was 144 in^2. What is the reduced ratio of the perimeters for these two rectangles?

End-of-Chapter Quiz

1. What is the sum of the interior angles of a 14-sided polygon?

   A. 1980°

   B. 2160°

   C. 2340°

   D. 2520°

2. If 4 interior angles of a pentagon are non-congruent obtuse angles, can the 5th angle be obtuse? Explain.

3. Determine the value of *x*.

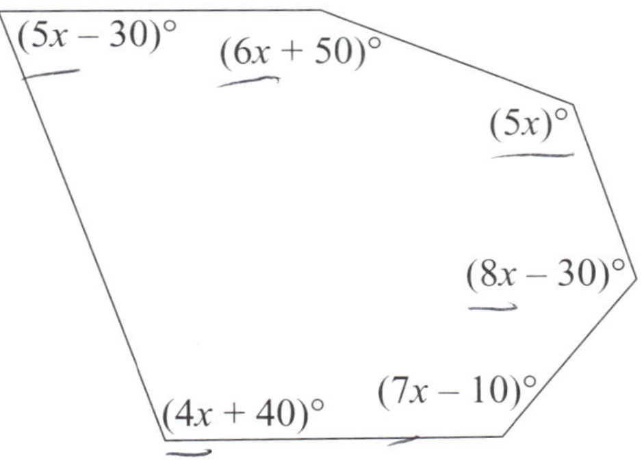

A. 10

B. 15

C. 20

D. 25

4. In a regular polygon, the measure of each exterior angle is half of the measure of each interior angle. How many sides does the polygon have?

A. 5

B. 6

C. 7

D. 8

5. ***ABCDE*** **≅** ***FGHIJ*****. Find the** $m\angle 1$**.**

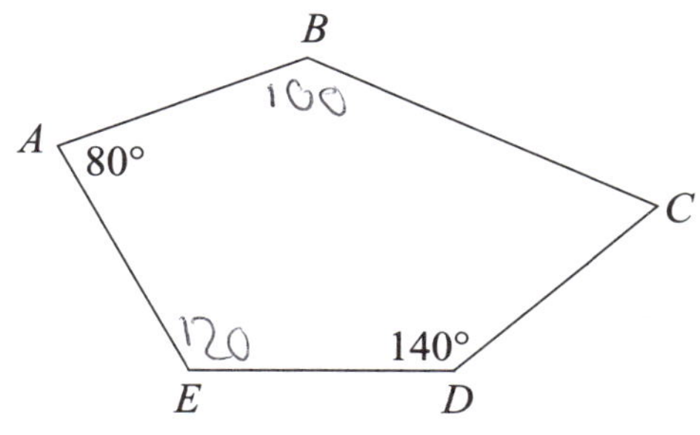

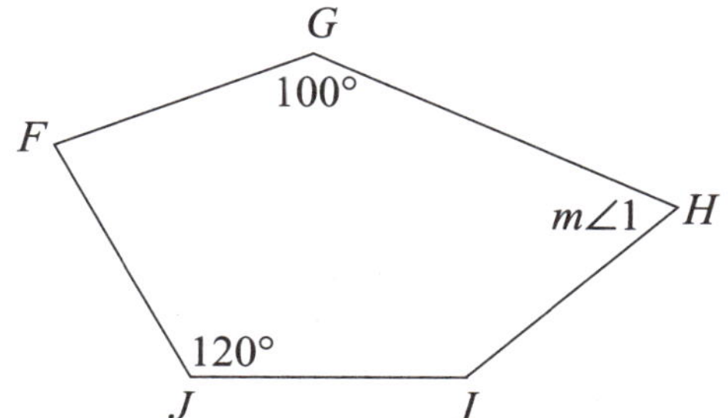

A. 80°

B. 100°

C. 120°

D. 140°

6. ***LMNO*** **~** ***PQRS*****. If the scale factor is 5 : 6, find the length of** $\overline{PS}$**.**

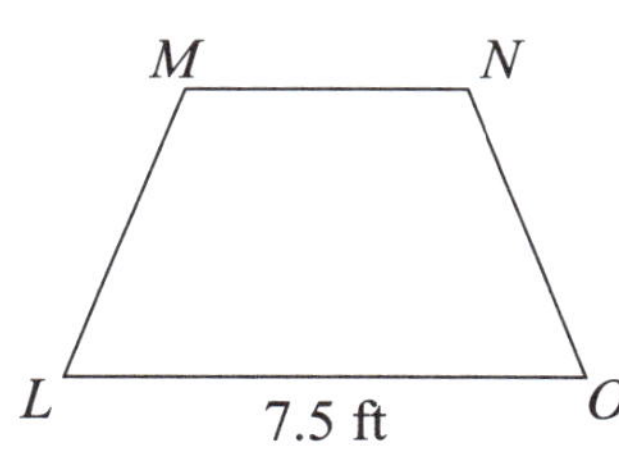

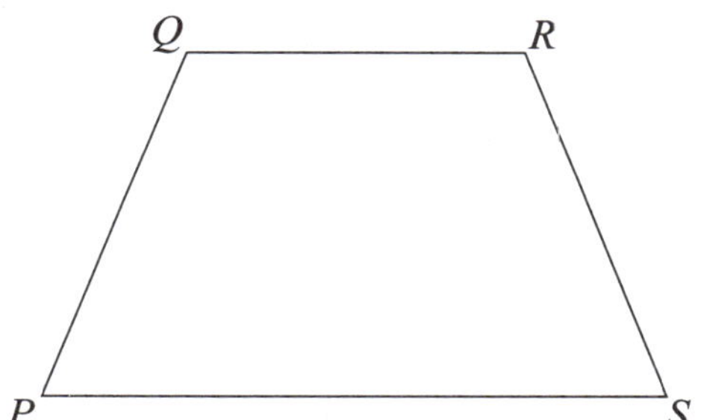

A. 6 ft

B. 8.5 ft

C. 9 ft

D. 12 ft

7. Prove that the rectangles are similar.

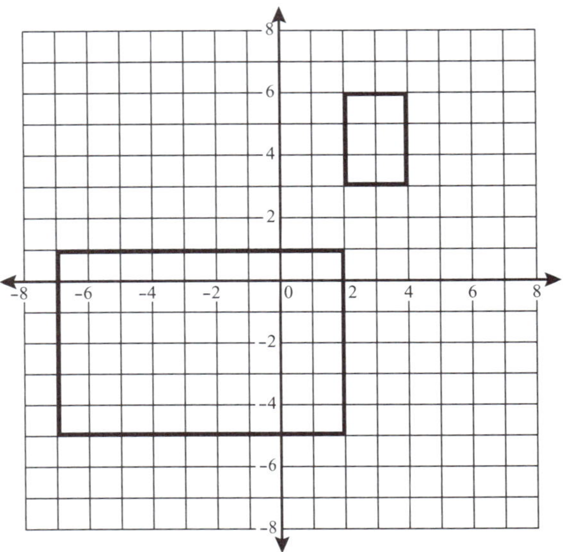

8. Which of the following describes a translation that will result in $\triangle A'B'C'$ being in quadrant 1?

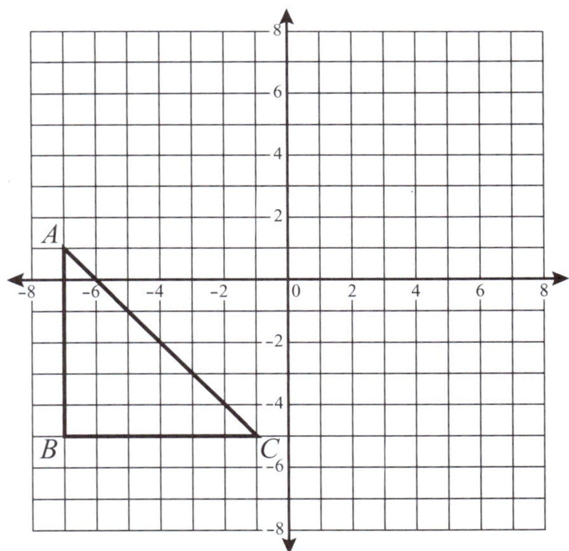

A. $(x, y) \rightarrow (x+6, y-6)$

B. $(x, y) \rightarrow (x-6, y-6)$

C. $(x, y) \rightarrow (x+8, y+8)$

D. $(x, y) \rightarrow (x-8, y+8)$

9. Marco stated that if $\triangle ABC$, from question 8, is reflected over the *y*-axis then the *x*-axis, it will result in the same image as reflecting $\triangle ABC$ first over the *x*-axis and then the *y*-axis. Is Marco correct? Explain.

10. Which of the following graphs shows $\overline{AB}$ rotated 270° counterclockwise about the origin?

A.

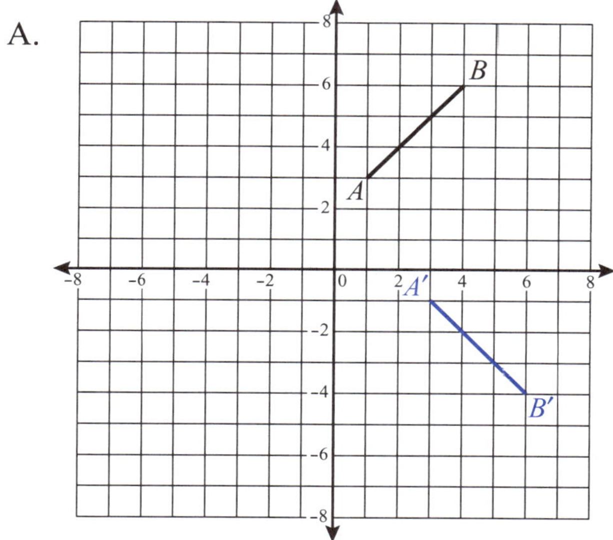

B.

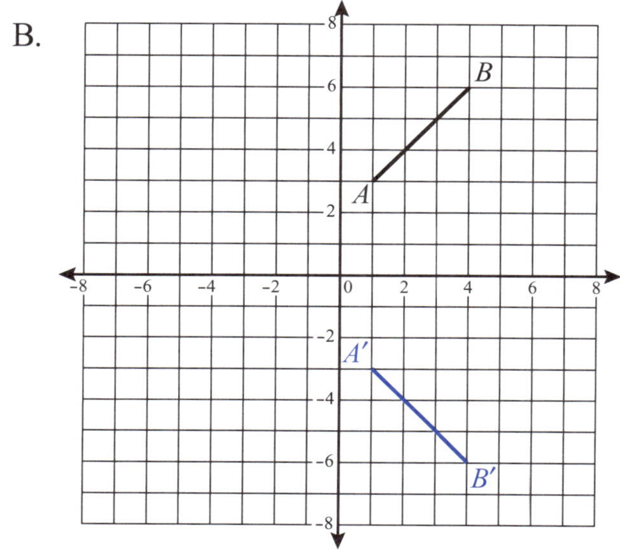

C.

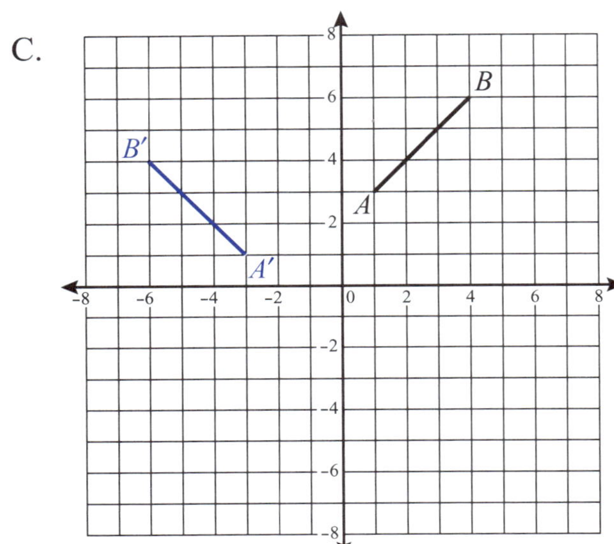

D.

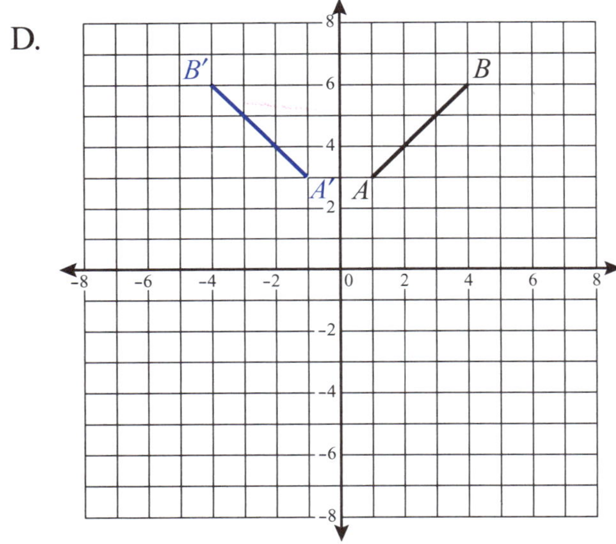

11. **Jonah is building an 8-inch by 11-inch picture frame out of wood. The wood frame is 2 inches wide on all sides. What is the area of the picture that fits in this frame?**

A. 28 in^2

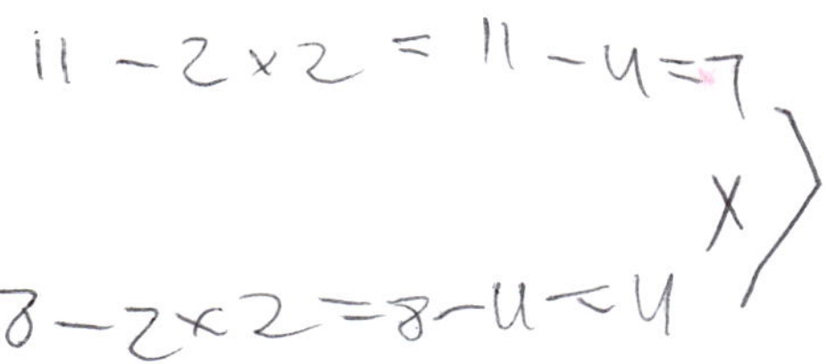

B. 36 in^2

C. 42 in^2

D. 54 in^2

12. **The area of the trapezoid below is 13 ft^2. What is the perimeter?**

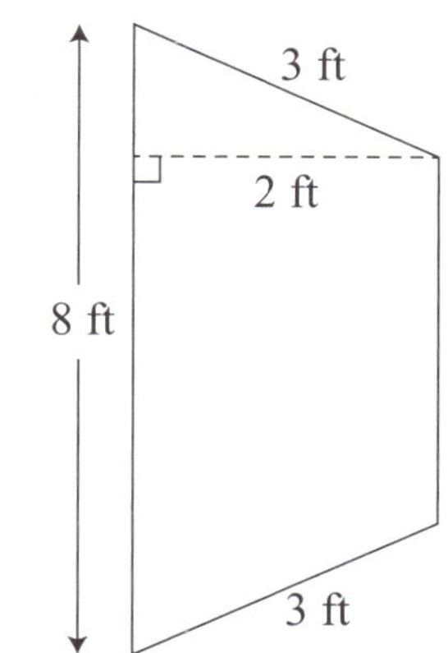

A. 16 ft

B. 19 ft

C. 21 ft

D. 24 ft

13. A regular pentagon has an area of 110 in^2. If the apothem is 5.5 inches, what is the length of each side?

A. 5 inches

B. 8 inches

C. 40 inches

D. Not enough information

14. The isosceles triangles below are similar. Find the ratio of their areas.

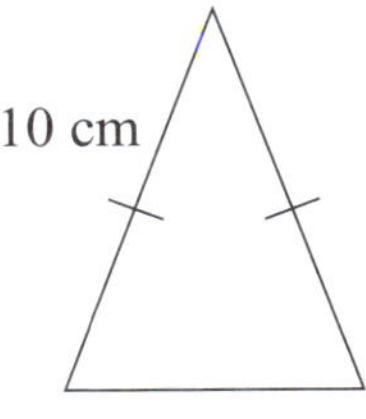

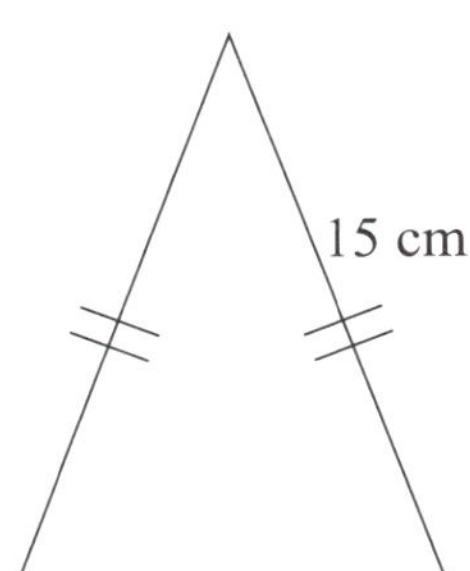

A. 2 : 3

B. 4 : 6

C. 4 : 9

D. 8 : 27

15. The areas of two similar rectangles are given below:

Rectangle X: Area = 27cm^2

Rectangle Y: Area = 48 cm^2

Which of the following are the dimensions for Rectangle *Y*?

A. 1 inch × 48 inches

B. 2 inches × 24 inches

C. 4 inches × 12 inches

D. 6 inches × 8 inches

Answers to Exercise 1

1. 2520°

$(n - 2)180°$

$(16 - 2)180°$

$(14)180° = 2520°$

2. 108°

$(n - 2)180°$

$(5 - 2)180°$

$(3)180° = 540°$

Since the polygon is regular, all 5 angles are congruent. Therefore, $m\angle 1 = 540 \div 5 = 108°$

3. 15

$(n - 2)180° = 2340°$

$$\frac{(n-2)180°}{180°} = \frac{2340°}{180°}$$

$n - 2 = 13$

$n - 2 + 2 = 13 + 2$

$n = 15$

4. 18

$360° \div 20° = 18$

5. 70°

$m\angle 1 + 110° = 180°$

$m\angle 1 + 110° - 110° = 180° - 110°$

$m\angle 1 = 70°$

Answers to Exercise 2

1. $\overline{AB} \cong \overline{WX}$

2. 85°

The figures are pentagons, so the sum of the interior angles = $(5 - 2)180° = 540°$

The figures are similar, so corresponding angles are congruent.

$m\angle 1 + 105° + 130° + 95° + 125° = 540°$

$m\angle 1 + 455° = 540°$

$m\angle 1 + 455° - 455° = 540° - 455°$

$m\angle 1 = 85°$

3. (–6, –2)

The figure slides 2 units left and 3 units up, so A' (–6, –2).

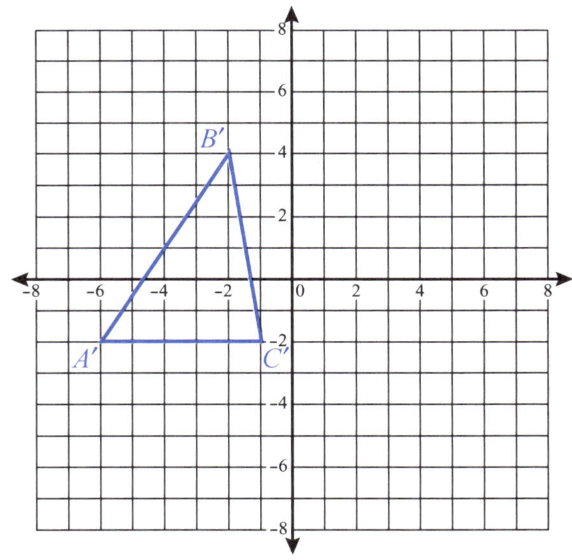

4.

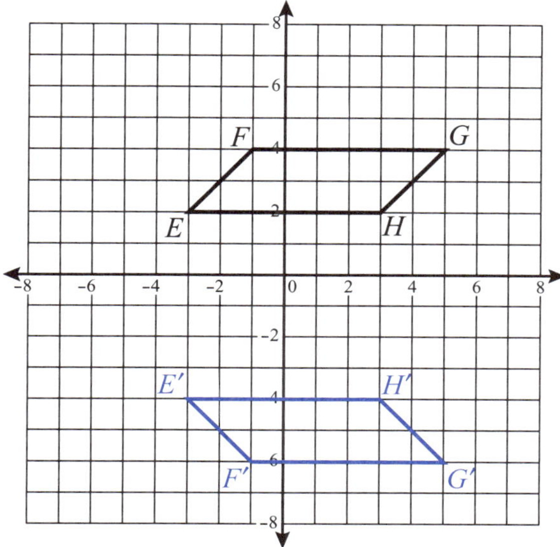

5. Scale factor = $\frac{1}{2}$

Answer to Exercise 3

1. 50 cm

$6^2 + 8^2 = C^2$

$36 + 64 = C^2$

$100 = C^2$

$\sqrt{100} = C^2$

$10 = C$

Perimeter = 10 cm + 10 cm + 15 cm + 15 cm

Perimeter = 50 cm

2. 105 m^2

$$A = \frac{1}{2}(\text{apothem})(\text{perimeter})$$

$$A = \frac{1}{2}(6)(42)$$

$$A = 126 \text{ m}^2$$

3. 12 ft = d_2

$$A = \frac{1}{2}d_1d_2$$

$$60 = \frac{1}{2}(10)d_2$$

$$60 = 5d_2$$

$$\frac{60}{5} = \frac{5d_2}{5}$$

$$12\text{ft} = d_2$$

4. ratio of areas = $8^2 : 11^2$ = 64 : 121

5. 5 : 6

100 : 144

$\sqrt{100} : \sqrt{144}$

10 : 12

5 : 6

Answers to the Chapter Quiz

1. B

$(n - 2)180°$

$(14 - 2)180°$

$(12)180° = 2160°$

2. Yes

The sum of the interior angles of a pentagon is 540° As an example, suppose that the other four angle measures are 91°, 92°, 93°, and 94°.

$m\angle 5 + 91° + 92° + 93° + 94° = 540°$

$m\angle 5 + 370° = 540°$

$m\angle 5 = 170°$

The 5th angle can be obtuse.

3. C

The sum of the interior angles of a hexagon is 720°

$(5x - 30) + (6x + 50) + (5x) + (8x - 30) + (7x - 10) + (4x + 40) = 720°$

$35x + 20 = 720$

$35x = 700$

$x = 20$

4. B

x: exterior angle

y: interior angle

$x + y = 180°$

$x = \frac{1}{2}y$

Substitute, $\frac{1}{2}y + y = 180°$

$\frac{3}{2}y = 180°$

$y = 120°$

$x = 60$

$360° \div 60° = 6$

5. B

$m\angle 1 + 140° + 120° + 80° + 100° = 540°$

$m\angle 1 + 440° = 540°$

$m\angle 1 = 100°$

6. C

$\frac{5}{6} = \frac{7.5}{x}$

$5x = 45$

$x = 9$

7. The small rectangle is 2 × 3
The large rectangle is 6 × 9

The ratio of corresponding sides is 2 : 6 and 3 : 9, which both simplify to 1 : 3. Therefore, the rectangles are similar.

8. C

The figure must move right and up to be in Quadrant 1.

9. Marco is correct.

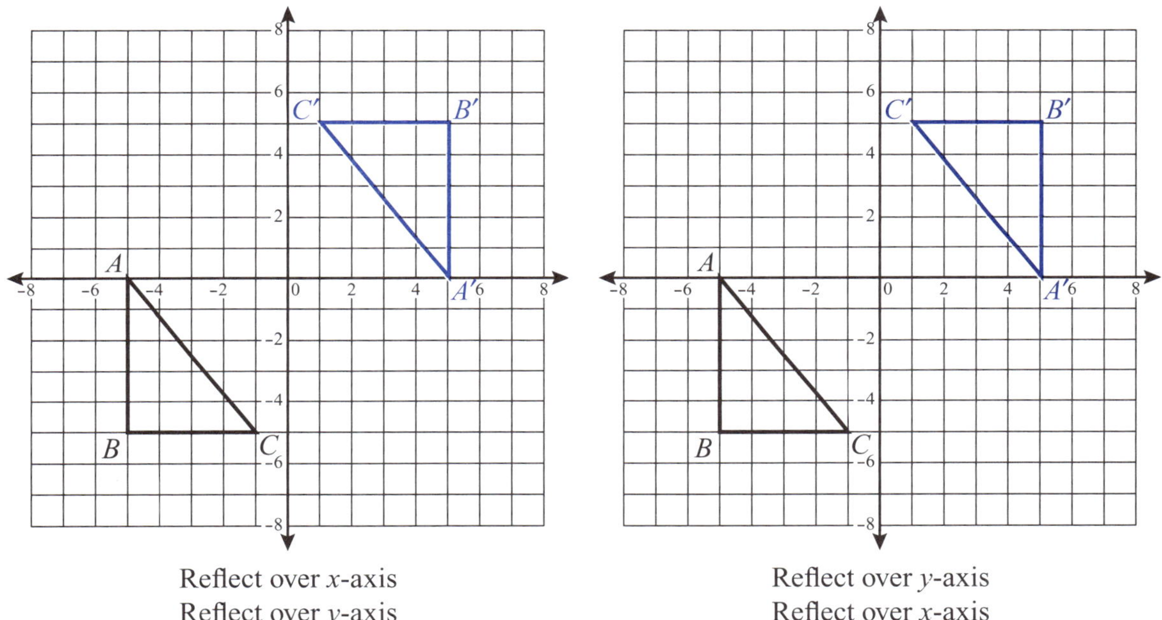

Reflect over x-axis
Reflect over y-axis

Reflect over y-axis
Reflect over x-axis

10. A

(1, 3) becomes (3, –1) and (4, 6) becomes (6, –4)

11. A

The frame is 2 inches on all sides, so the length and width of the picture are each 4 inches less.

$A = (4)(7) = 28 \text{ in}^2$

12. B

$$A = \frac{1}{2}h(b_1 + b_2)$$

$$13 = \frac{1}{2}(2)(8 + b_2)$$

$$13 = 1(8 + b_2)$$

$$13 = 8 + b_2$$

$$5 = b_2$$

Perimeter = 8 + 3 + 3 + 5 = 19 ft

13. B

$$A = \frac{1}{2}(\text{apothem})(\text{perimeter})$$

$$110 = \frac{1}{2}(5.5)(\text{perimeter})$$

$$110 = 2.75(\text{perimeter})$$

$$40 = \text{perimeter}$$

Since the polygon is a regular pentagon, divide the perimeter by 5.

$40 \div 5 = 8$ inches

14. C

$a : b$	so,	$a^2 : b^2$
10 : 15		4 : 9
2 : 3		

15. C

The ratio of the areas is 27 : 48, which simplifies to 9 : 16.

Since $a^2 : b^2$ is 9 : 16, the ratio of the corresponding sides is 3 : 4.

Rectangle X Rectangle Y

3 4

Thus, given their areas:

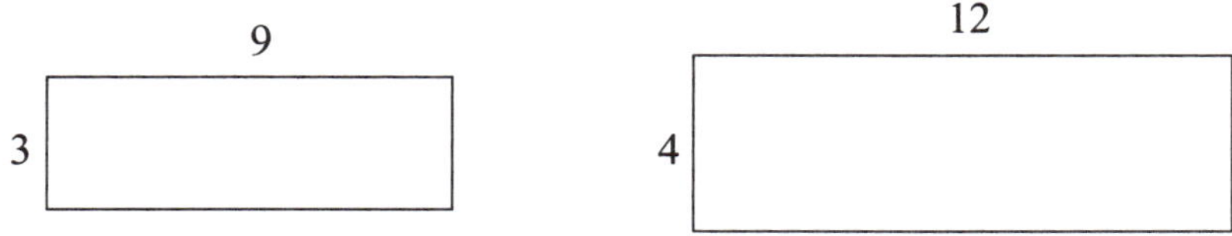

Rectangles X and Y have these dimensions.

Chapter 3

Quadrilaterals

Your Goals for Chapter 3

1. You should know the properties of quadrilaterals and use them to solve problems.
2. You should be able to use coordinate geometry to prove the properties and theorems of special quadrilaterals.

Standards

The following standards are assessed on Florida's Geometry End-of-Course exam either directly or indirectly:

MA.912.G.3.1: (Moderate) Describe, classify, and compare relationships among quadrilaterals including the square, rectangle, rhombus, parallelogram, trapezoid, and kite.

MA.912.G.3.2: (Moderate) Compare and contrast special quadrilaterals on the basis of their properties.

MA.912.G.3.3: (High) Use coordinate geometry to prove properties of congruent, regular, and similar quadrilaterals.

MA.912.G.3.4: (High) Prove theorems involving quadrilaterals.

Special Quadrilaterals

A **quadrilateral** is a polygon with four sides. There are several types of special quadrilaterals.

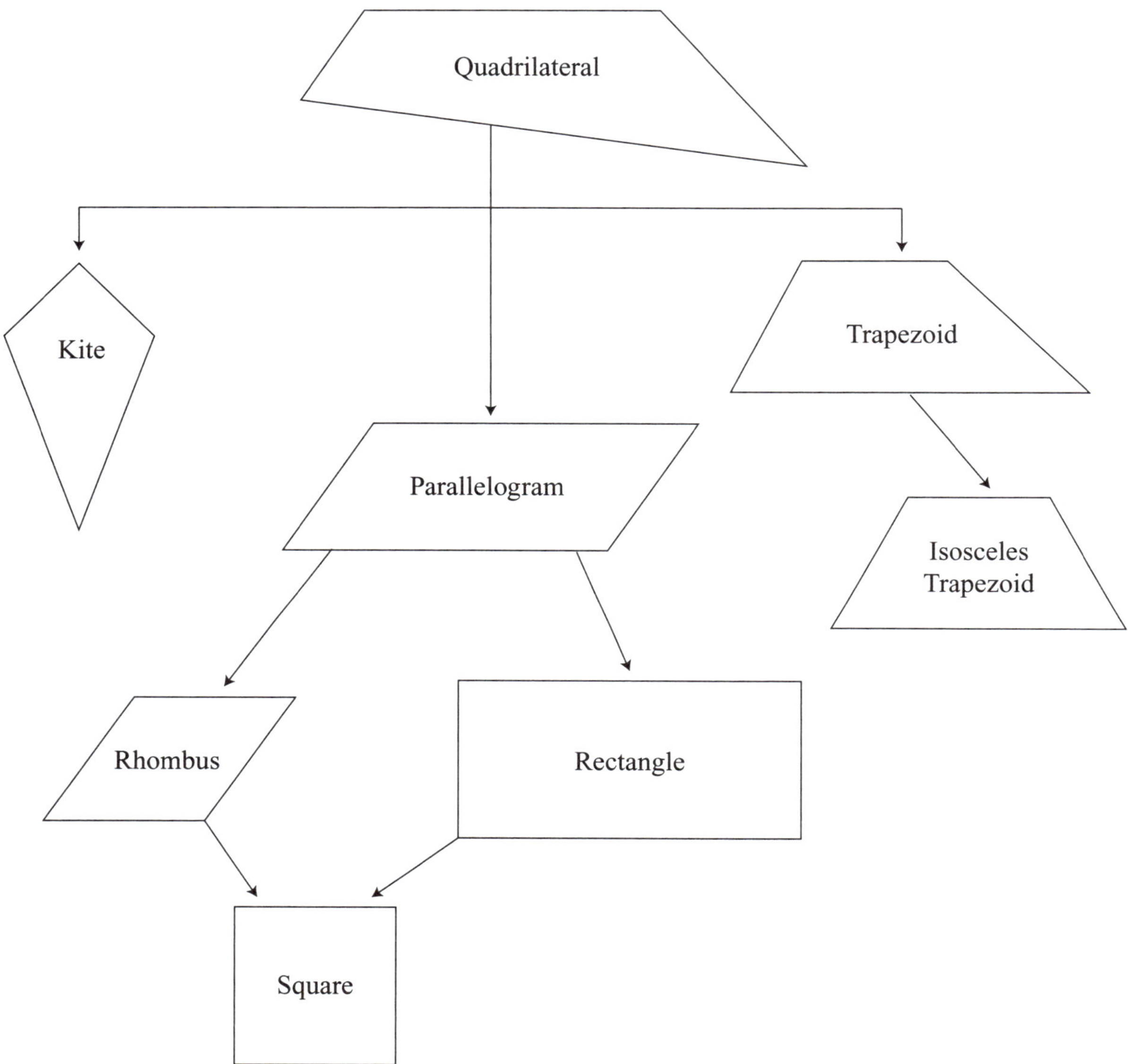

The above figure looks like a tree diagram. It can be thought of as the "Quadrilateral Family Tree" because it shows the relationships among the different types of special quadrilaterals.

Example:

True or False.

a. A square is a type of rectangle.

b. A parallelogram is a type of rhombus.

c. A rhombus is a type of kite.

d. A parallelogram is a type of trapezoid.

e. A rectangle is a type of rhombus.

a. True: Although a square does have more specific properties than a rectangle, it does maintain all of the properties of a rectangle. Therefore, a square is a rectangle.

b. False: A rhombus must have four congruent sides and a parallelogram does not. Therefore, a parallelogram cannot be a type of rhombus. (However, a rhombus is a type of parallelogram.)

c. False: A kite has two separate pairs of congruent adjacent sides rather than the four congruent sides of a rhombus.

d. False: A trapezoid can have only one pair of parallel sides. A parallelogram must have two pairs of parallel sides.

e. False: A rectangle does not have four congruent sides, therefore, it cannot be a rhombus.

Properties of Quadrilaterals

Parallelogram

A quadrilateral with both pairs of opposite sides parallel.

Properties:

- both pairs of opposite sides are congruent
- consecutive angles are supplementary
- both pairs of opposite angles are congruent
- diagonals bisect each other

Rectangle

A type of parallelogram with four right angles.

Properties:

- both pairs of opposite sides are congruent
- all angles are right angles
- diagonals bisect each other
- both pairs of opposite sides are parallel
- diagonals are congruent
- consecutive angles are supplementary

Rhombus

A type of parallelogram with all sides congruent.

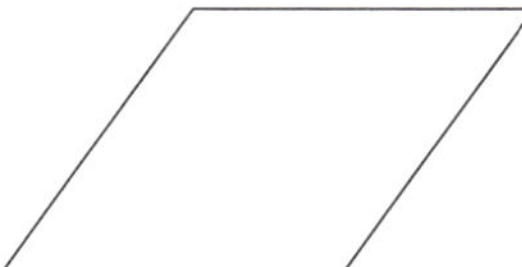

Properties:

- all sides are congruent
- both pairs of opposite angles are congruent
- diagonals are perpendicular bisectors
- both pairs of opposite sides are parallel
- consecutive angles are supplementary
- diagonals bisect the angles

Square

A type of parallelogram with all sides congruent and four right angles.

Properties:

- all sides are congruent
- all angles are right angles
- diagonals are perpendicular bisectors
- diagonals bisect the angles
- opposite sides are parallel
- diagonals are congruent
- consecutive angles are supplementary

Trapezoid

A quadrilateral with exactly one pair of parallel sides.

Properties:

- consecutive angles along non-parallel sides are supplementary
- only one pair of sides is parallel

Isosceles Trapezoid

A type of trapezoid with congruent base angles and congruent non-parallel sides.

Properties:

- non-parallel sides are congruent
- diagonals are congruent
- congruent base angles
- only one pair of sides is parallel
- base angles are congruent
- opposite angles are supplementary
- consecutive angles along non-parallel sides are supplementary

Kite

A quadrilateral with two distinct pairs of adjacent sides that are congruent.

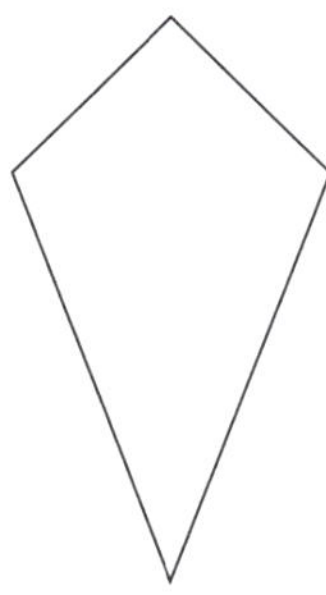

Properties:

- diagonals are perpendicular
- exactly one pair of opposite angles is congruent
- there are two distinct pairs of congruent adjacent sides

Note: Opposite sides are neither parallel nor congruent.

Example:

What kind of quadrilateral is *PQRS*?

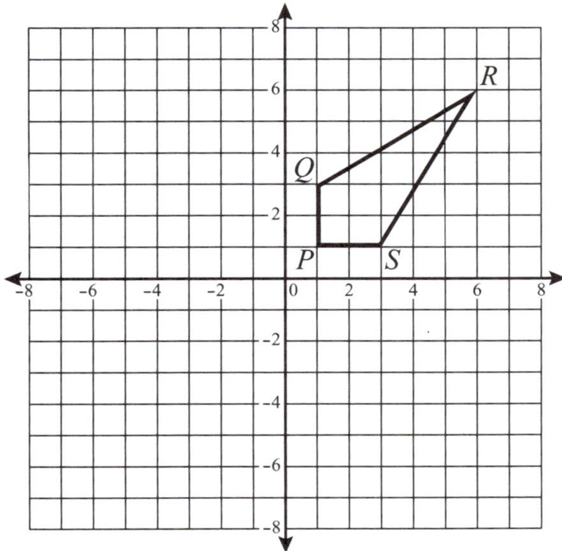

$P(1, 1)$, $Q(1, 3)$, $R(6, 6)$, $S(3, 1)$

Plot the points and connect to form a figure.

$PQ = 2$ units and $PS = 2$ units, therefore $\overline{PQ} \cong \overline{PS}$.

$QR = \sqrt{(6-1)^2 + (6-3)^2} = \sqrt{34}$ and $SR = \sqrt{(6-3)^2 + (6-1)^2} = \sqrt{34}$, therefore $\overline{QR} \cong \overline{SR}$.

Since there are two pairs of congruent adjacent sides, but opposite sides are not congruent, the figure is a kite.

PQRS is a kite.

Example:

Given a quadrilateral below, determine if the diagonals are perpendicular.

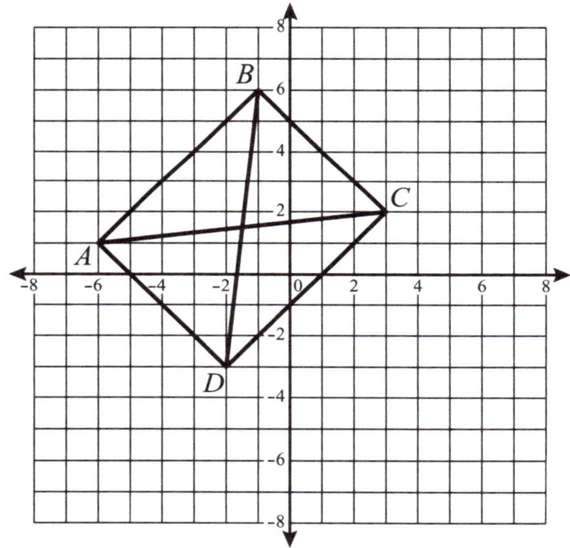

Lines are perpendicular if their slopes are opposite reciprocals.

$\overline{AC}$ passes through the points (–6, 1) and (3, 2).

The slope of the line $= \dfrac{2-1}{3-(-6)} = \dfrac{1}{9}$

$\overline{BD}$ passes through the points (–1, 6) and (–2, –3).

The slope of the line $= \dfrac{-3-6}{-2-(-1)} = \dfrac{-9}{-1} = 9$

The slopes are reciprocals, but are not opposite signs. Therefore, the diagonals are not perpendicular.

Example:

Figure *LMNO* is a square. The length of $\overline{LP} = 3x + 5$ and the length of $\overline{PN} = x + 11$. Determine the length of $\overline{LN}$.

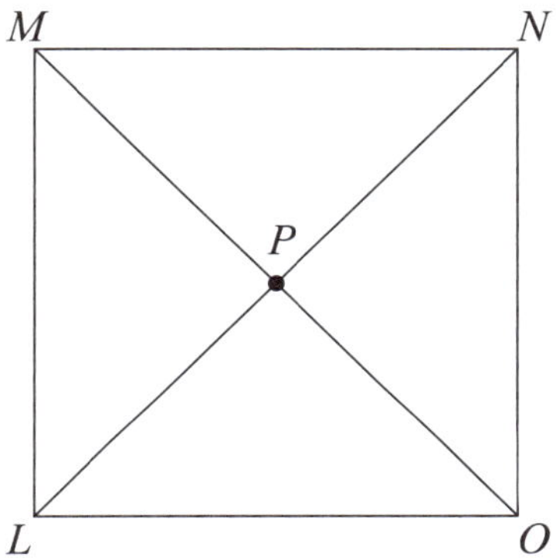

The diagonals of a square bisect each other. Therefore,

$3x + 5 = x + 11$

$3x - x + 5 = x - x + 11$ Subtract x from both sides

$2x + 5 = 11$

$2x + 5 - 5 = 11 - 5$ Subtract 5 from both sides

$2x = 6$

$\dfrac{2x}{2} = \dfrac{6}{2}$ Divide both sides by 2

$x = 3$

The length of $\overline{LN} = 3x + 5 + x + 11$.

Substitute $x = 3$ into the equation.

$\overline{LN} = 3(3) + 5 + 3 + 11$

$\overline{LN} = 28$

End-of-Chapter Quiz

1. Which of the following quadrilaterals has diagonals that are perpendicular bisectors?

A. Rectangle

B. Square

C. Rhombus

D. Both rectangle and square

E. Both square and rhombus

2. Which of the following is a type of rhombus?

A. Kite

B. Rectangle

C. Square

D. Trapezoid

3. If figure *EFGH* is a rectangle, which of the following is *not* true?

A. Opposite sides are congruent

B. Consecutive angles are supplementary

C. Diagonals are congruent

D. Diagonals are perpendicular

4. Figure *WXYZ* is a trapezoid. If $\angle X = (20a + 10)°$ and $\angle W = (15a - 5)°$, what is $m\angle X$?

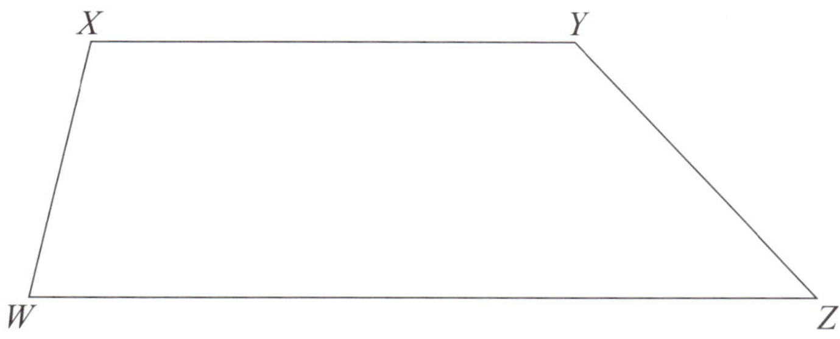

A. 40°

B. 70°

C. 110°

D. 180°

Use the information in the parallelogram to answer questions 5–8.

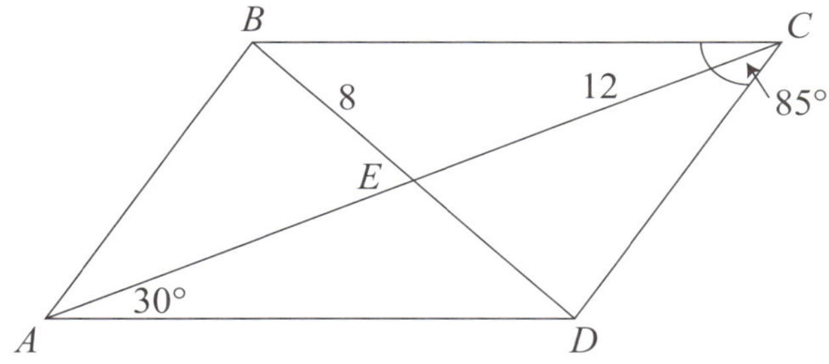

5. Find $m\angle DAB$.

A. 30°

B. 60°

C. 85°

D. 95°

6. Find $m\angle CAB$.

A. 30°

B. 55°

C. 85°

D. 90°

7. Find *DE*.

A. 8

B. 12

C. 16

D. 24

8. Find *AC*.

A. 8

B. 12

C. 16

D. 24

Use the information in the rhombus to answer questions 9–12.

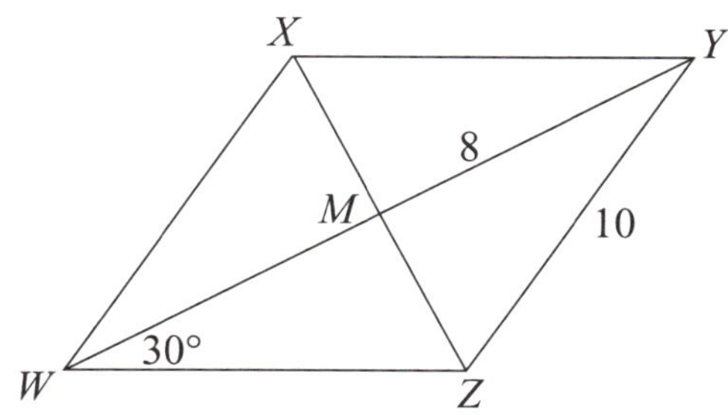

9. Find *XY*.

A. 6

B. 8

C. 10

D. 18

10. Find *MZ*.

A. 6

B. 8

C. 10

D. 18

11. Find $m\angle XMY$.

A. 30°

B. 60°

C. 90°

D. 120°

12. Find $m\angle XWZ$.

A. 30°

B. 60°

C. 90°

D. 120°

Use the information in the isosceles trapezoid to answer questions 13 and 14.

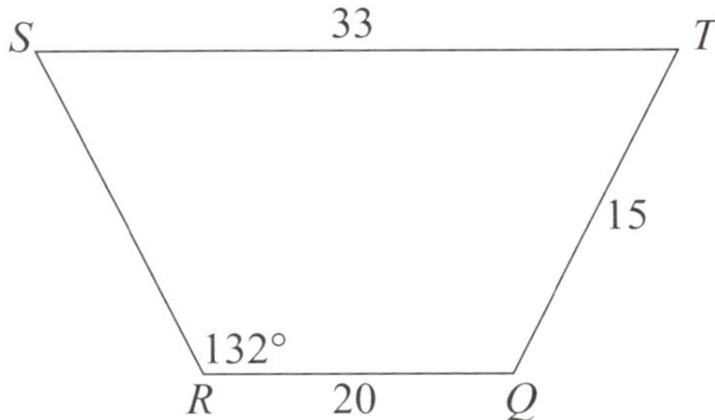

13. Find the perimeter of the isosceles trapezoid.

A. 68

B. 83

C. 88

D. 96

14. Find $m\angle STQ$.

A. 48°

B. 66°

C. 132°

D. Not enough information

15. Figure *ABCD* has vertices (1, 1), (1, 5), (–6, 1), and (–6, 5). Prove that the diagonals are congruent.

Answers to the Chapter Quiz

1. E

Both a square and a rhombus have diagonals that are perpendicular bisectors.

2. C

A square is a type of rhombus because it has all the properties of a rhombus.

3. D

The diagonals of a rectangle are not perpendicular.

4. C

$\angle X$ and $\angle W$ are supplementary. Therefore,

$$(20a+10)^\circ + (15a-5)^\circ = 180^\circ$$

$$35a+5 = 180^\circ$$

$$35a = 175^\circ$$

$$a = 5$$

Substitute $a = 5$ into $\angle X = (20a + 10)^\circ$.

$\angle X = (20(5) + 10)^\circ = 110^\circ$

5. C

Opposite angles of a parallelogram are congruent.

6. B

$m\angle DAB = m\angle DAC + m\angle CAB$

$85^\circ = 30^\circ + m\angle CAB$

$55^\circ = m\angle CAB$

7. A

Diagonals of a parallelogram bisect each other, so $DE = EB$.

8. D

$AC = AE + EC$

Diagonals of a parallelogram bisect each other, so $AE = EC$.

$AC = 12 + 12 = 24$

9. C

All sides of a rhombus are congruent.

10. A

Since the diagonals of a rhombus are perpendicular, $m\angle YMZ = 90°$. Therefore, $\triangle YMZ$ is a right triangle and the Pythagorean Theorem can be used to find MZ.

$(MZ)^2 + 8^2 = 10^2$

$(MZ)^2 + 64 = 100$

$(MZ)^2 = 36$

$\sqrt{(MZ)^2} = \sqrt{36}$

$MZ = 6$

11. C

The diagonals of a rhombus are perpendicular.

12. B

The diagonals of a rhombus bisect opposite angles.

13. B

Since the trapezoid is isosceles, $RS = QT = 15$. To find the perimeter, add all sides.

$15 + 15 + 20 + 33 = 83$

14. A

Consecutive angles along non-parallel sides of a trapezoid are supplementary.

$m\angle SRQ + m\angle TSR = 180^\circ$

$132^\circ + m\angle TSR = 180^\circ$

$m\angle TSR = 48^\circ$

Base angles of an isosceles triangle are congruent.

$m\angle TSR = m\angle STQ$

$48^\circ = m\angle STQ$

15. Use the distance formula to show that the diagonals of the rectangle are congruent.

Diagonal 1

$d = \sqrt{(x_2 - x_1)^2 + (y_2 - y_1)^2}$

$d = \sqrt{(-6-1)^2 + (1-5)^2}$

$d = \sqrt{(-7)^2 + (-4)^2}$

$d = \sqrt{49+16}$

$d = \sqrt{65}$

Diagonal 2

$d = \sqrt{(x_2 - x_1)^2 + (y_2 - y_1)^2}$

$d = \sqrt{(-6-1)^2 + (5-1)^2}$

$d = \sqrt{(-7)^2 + (4)^2}$

$d = \sqrt{49+16}$

$d = \sqrt{65}$

Diagonals are congruent.

Chapter 4

Triangles

Your Goals for Chapter 4

1. You should be able to classify triangles.
2. You should be able to define, identify and construct geometric terminology associated with triangles.
3. You should be able to prove that triangles are congruent or similar and use their properties to solve problems.
4. You should be able to apply theorems involving segments divided proportionally.
5. You should be able to apply the triangle inequality theorems.

Standards

The following standards are assessed on Florida's Geometry End-of-Course exam either directly or indirectly:

MA.912.G.4.1: (Moderate) Classify, construct, and describe triangles that are right, acute, obtuse, scalene, isosceles, equilateral, and equiangular.

MA.912.G.4.2: (Moderate) Define, identify, and construct altitudes, medians, angle bisectors, perpendicular bisectors,orthocenter, centroid, incenter, and circumcenter.

MA.912.G.4.3: (High) Construct triangles congruent to given triangles.

MA.912.G.4.4: (Moderate) Use properties of congruent and similar triangles to solve problems involving lengths and areas.

MA.912.G.4.5: (Moderate) Apply theorems involving segments divided proportionally.

MA.912.G.4.6: (High) Prove that triangles are congruent or similar and use the concept of corresponding parts of congruent triangles.

MA.912.G.4.7: (Moderate) Apply the inequality theorems: triangle inequality, inequality in one triangle, and the "Hinge Theorem."

Classification

Triangles are classified by the length of their sides and the measure of their angles.

Classification by Side Length	
Equilateral	3 congruent sides
Isosceles	2 congruent sides
Scalene	no congruent sides

Classification by Angle Measure	
Acute	3 acute angles
Right	1 right angle
Obtuse	1 obtuse angle
Equiangular	3 congruent angles

Example:

Troy's teacher asked him to construct an isosceles right triangle. What must be the measure of the three angles?

By definition, an isosceles right triangle has one right angle. Therefore, the measure of one angle is 90°. Since the triangle is also isosceles, it has two congruent sides. If a triangle has two congruent sides, it also has two congruent angles. The sum of the degrees in a triangle is 180°. With one angle measuring 90°, there are 90° remaining to be split evenly between the two remaining congruent angles.

$$\frac{90°}{2} = 45°$$

The three angles are 90°, 45°, and 45°.

Example:

Classify the given triangle by angle measure.

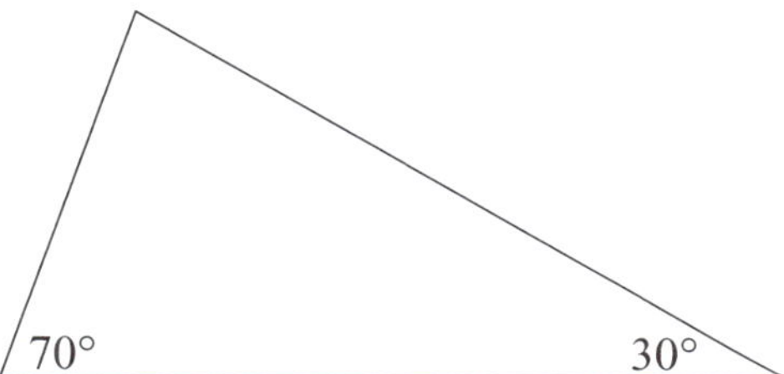

Both angles given are acute. The third angle measures 80° because the sum of the degrees of a triangle is 180°. Since all three angles are acute, the triangle can be classified as acute.

Definitions Related to Triangles

An **altitude of a triangle** is a perpendicular segment from a vertex to its opposite side. Every triangle has three altitudes, which are also referred to as the height of the triangle.

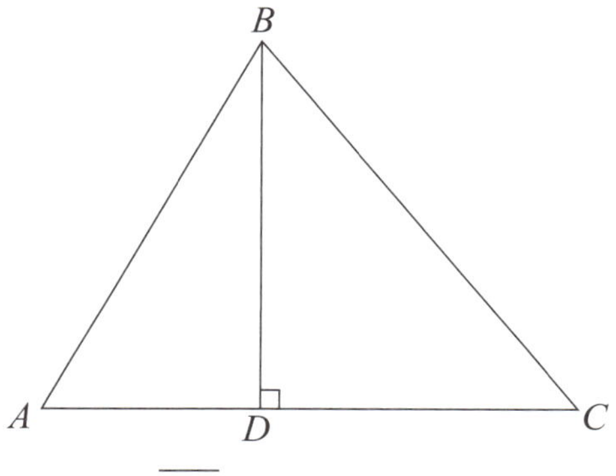

$\overline{BD}$ is an altitude

A **median of a triangle** is a line segment joining a vertex to the midpoint of its opposite side. Every triangle has three medians.

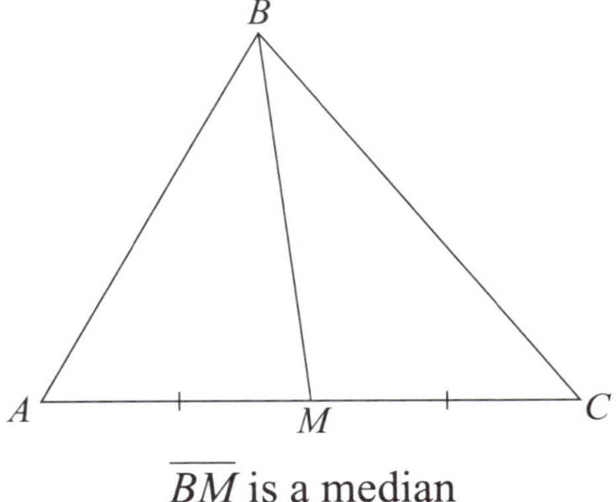

$\overline{BM}$ is a median

An **angle bisector** in a triangle is a segment that is drawn from a vertex and cuts the angle in half. Every triangle has three angle bisectors.

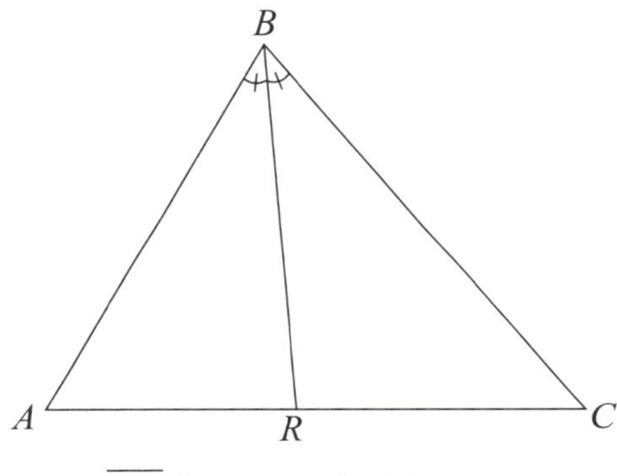

$\overline{BR}$ is an angle bisector

A **perpendicular bisector** of a triangle is a line that passes through the midpoint of a side and is perpendicular to that given side. Every triangle has three perpendicular bisectors.

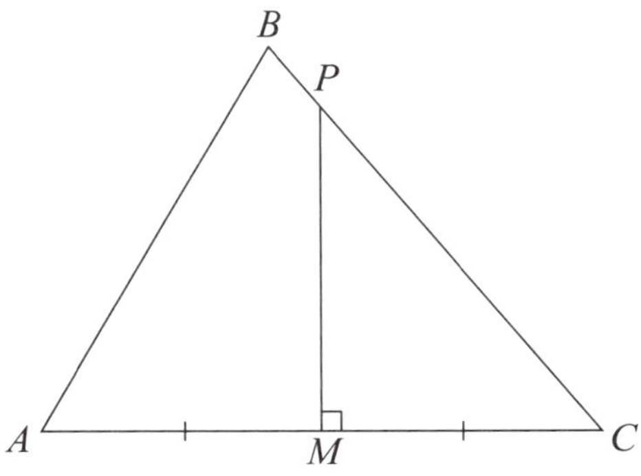

$\overline{PM}$ is a perpendicular bisector

Points of Concurrency

When three or more lines or segments intersect at the same point, the lines are called **concurrent lines**.

The point at which concurrent lines intersect is called the **point of concurrency**.

circumcenter of a triangle: the point of concurrency of the 3 perpendicular bisectors of a triangle. The circumcenter can be inside, on or outside the triangle.

The circumcenter is equidistant from the vertices of a triangle.

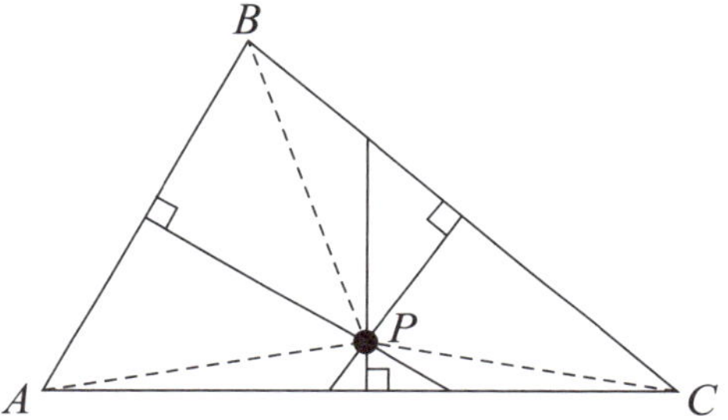

P is the circumcenter

$PA = PB = PC$

incenter of a triangle: the point of concurrency of the 3 angle bisectors of a triangle. The incenter is always inside the triangle.

The incenter is equidistant from the 3 sides of a triangle.

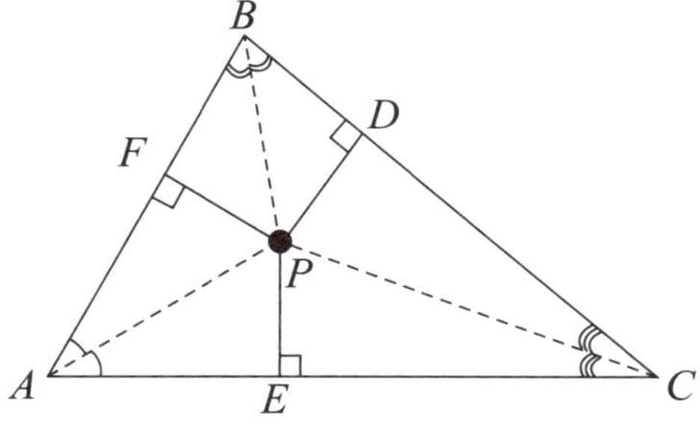

P is the incenter

$PD = PE = PF$

centroid of a triangle: the point of concurrency of the 3 medians of a triangle. The centroid is always inside the triangle. (**median:** a segment whose endpoints are a vertex of a triangle and the midpoint of the opposite side)

The centroid is two thirds of the distance from each vertex to the midpoint of the opposite side.

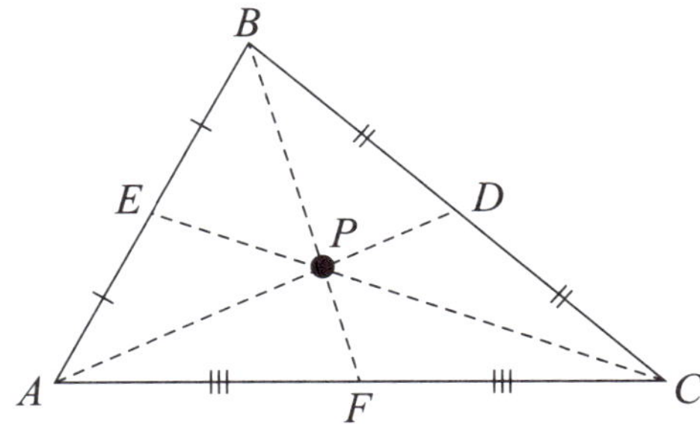

P is the centroid

$$AP = \frac{2}{3}AD,\ BP = \frac{2}{3}BF,\ CP = \frac{2}{3}CE$$

orthocenter of a triangle: the point of concurrency of the 3 altitudes of a triangle. The orthocenter can occur inside, on or outside a triangle.

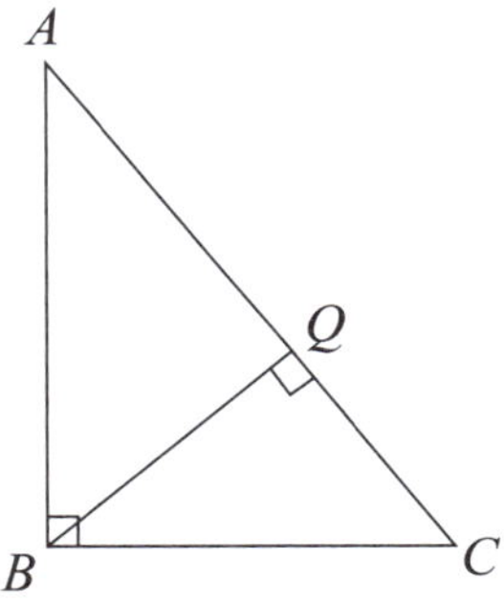

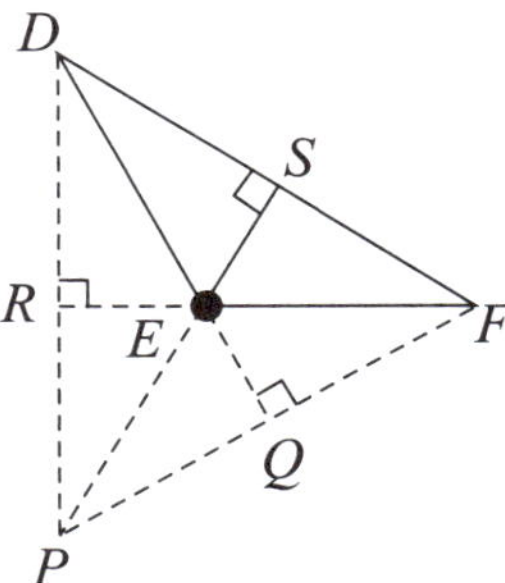

Although the orthocenter, *P*, is shown to lie inside $\triangle ABC$, it can be located on a vertex of a triangle or outside the triangle. These two situations are shown above.

The figure on the left is a right triangle, with $m\angle ABC = 90°$. The three altitudes are $\overline{AB}$, $\overline{CB}$, $\overline{BQ}$. The orthocenter is point *B*. The figure on the right is an obtuse triangle, with $m\angle DEF > 90°$. The three altitudes are $\overline{DR}$, $\overline{FQ}$, $\overline{ES}$. The orthocenter *P* lies outside the triangle.

Example:

P is the centroid of $\triangle ABC$ and BP = 10. Find PD and BD.

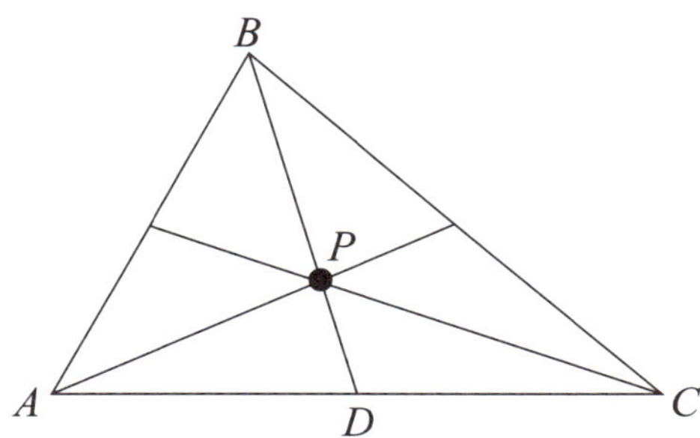

Since P is the centroid, $BP = \frac{2}{3}BD$.

Using substitution,

$$10 = \frac{2}{3}BD$$

$$15 = BD.$$

$$PD = BD - BP$$

$$PD = 15 - 10 = 5$$

$$PD = 5$$

Example:

Point P is the incenter of $\triangle ABC$.

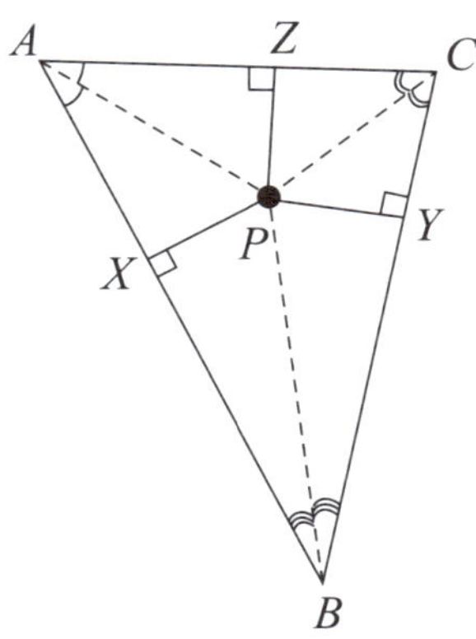

Given $AX = 15$ and $AP = 17$, find PX and PZ.

Using the Pythagorean Theorem to find PX in $\triangle APX$.

$(PX)^2 + (AX)^2 = (AP)^2$

$(PX)^2 + 15^2 = 17^2$

$(PX)^2 + 225 = 289$

$(PX)^2 = 64$

$PX = 8$

Since the incenter is equidistant from all three sides, $PX = PZ$.

$PX = PZ = 8$

Exercise 1

1. **Jackson stated that for a triangle to be obtuse, all three angles must be obtuse. Is this possible? Explain.**

2. **If *Z* is a midpoint and *Y* is a centroid, find the length of $\overline{XZ}$.**

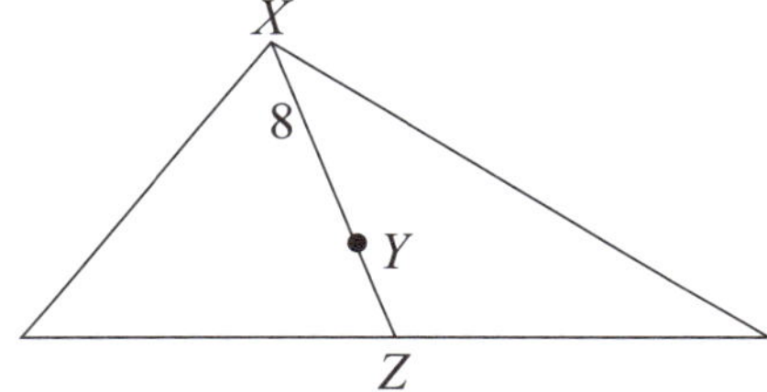

3. **Find the circumcenter of $\triangle ABC$ with vertices $A(-4, -2)$, $B(0, 2)$, and $C(4, -2)$.**

Congruent Triangles

Triangles are congruent when all corresponding sides and corresponding angles are congruent. The triangles will be the same shape and the same size.

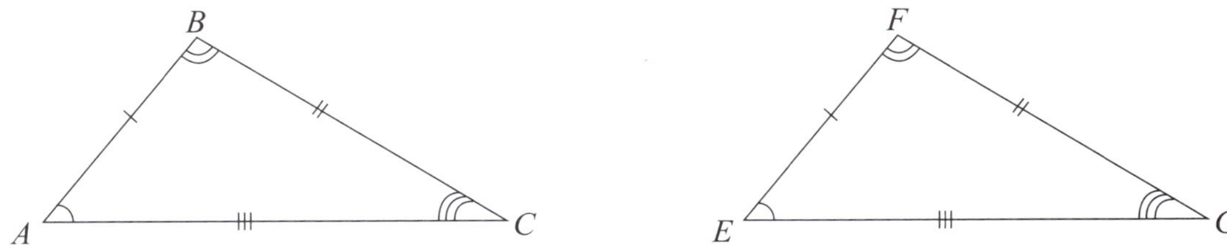

$\triangle ABC \cong \triangle EFG$ because,

corresponding sides	corresponding angles
$AB = EF$	$\angle A = \angle E$
$AC = EG$	$\angle B = \angle F$
$BC = FG$	$\angle C = \angle G$

A triangle is defined by a set of six measurements: three sides and three angles. For a pair of triangles, certain groups of three of these measurements can prove their congruency. Triangles are congruent if:

SSS (side side side): All three pairs of corresponding sides are equal.

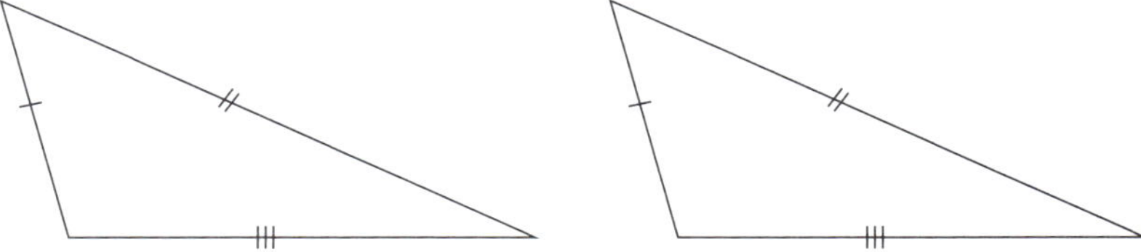

SAS (side angle side): Two pairs of corresponding sides and their included angle are equal.

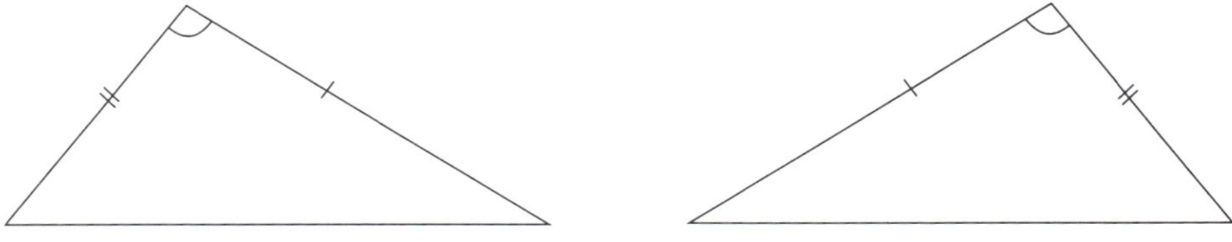

ASA (angle side angle): Two pairs of corresponding angles and their included side are equal.

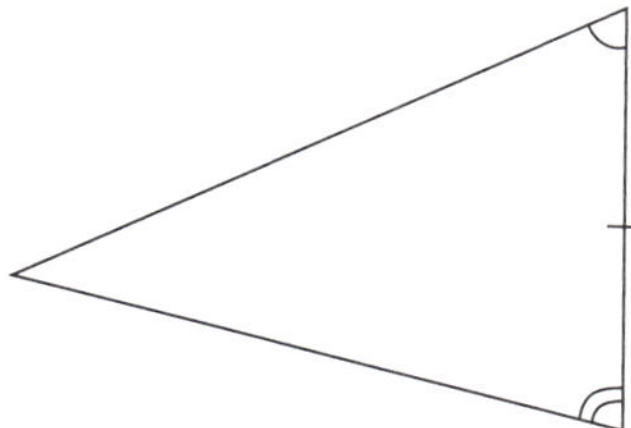
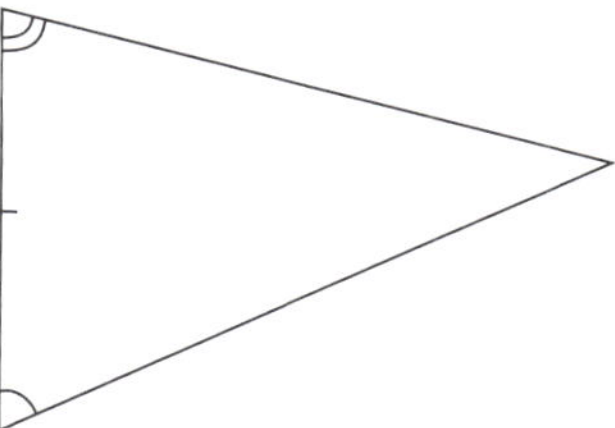

AAS (angle angle side): Two pairs of corresponding angles and a pair of non-included sides are equal.

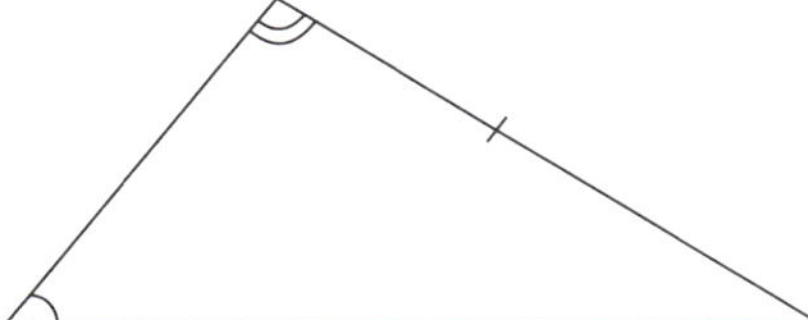
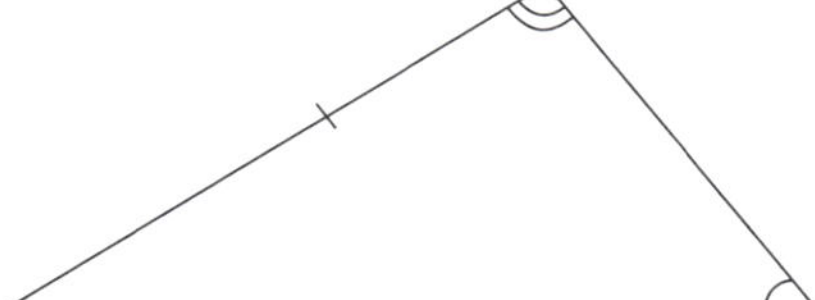

HL (hypotenuse and leg of a right triangle): The hypotenuses and one pair of corresponding legs of two right triangles are equal.

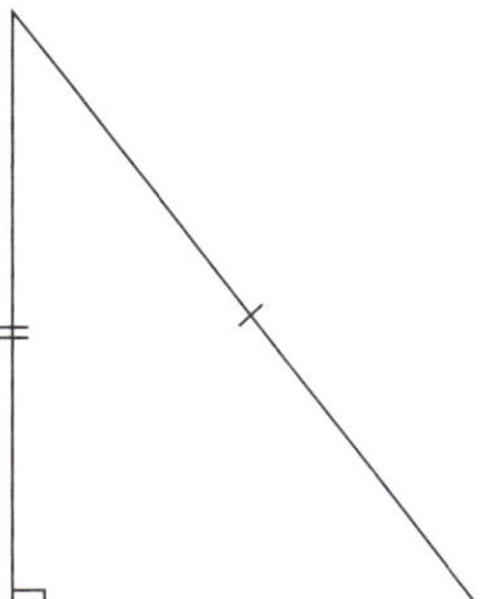
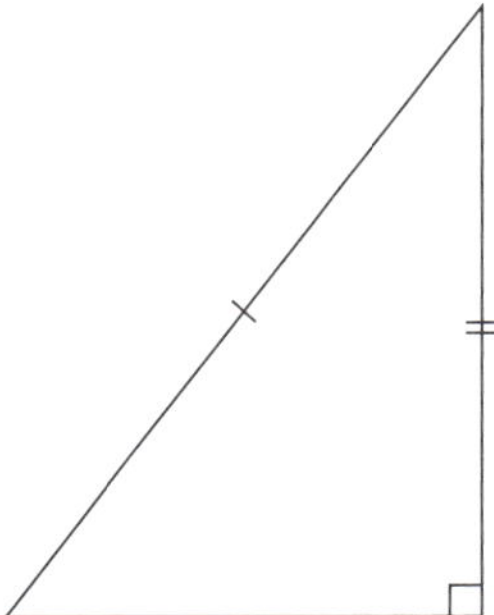

Example:

Determine if the triangles are congruent by SSS, SAS, ASA, AAS, HL, or none of these.

a.

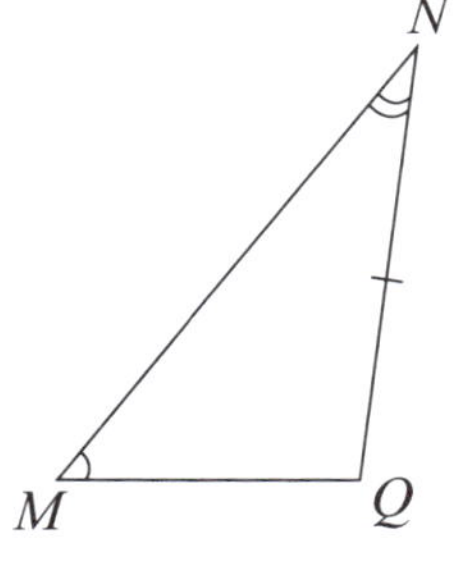

b.

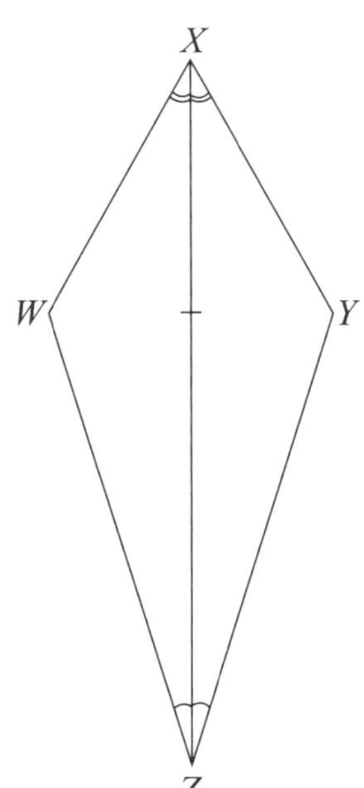

c.

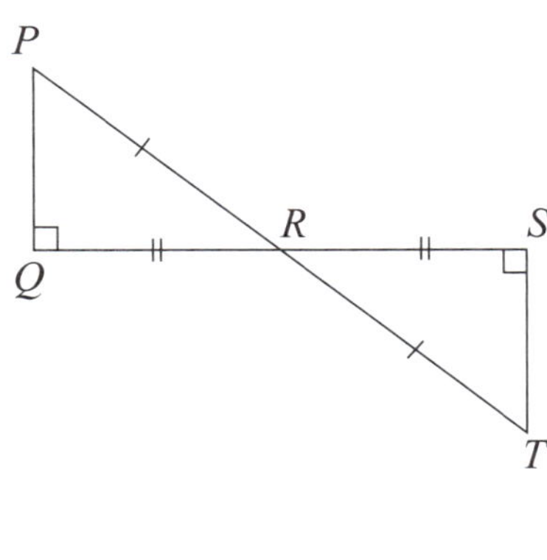

a. Two pairs of corresponding angles are equal.
$\angle M = \angle A$
$\angle N = \angle B$
A pair of non-included corresponding sides are equal.
$QN = CB$
AAS

b. Two pairs of corresponding angles are equal.
$\angle WXZ = \angle YXZ$
$\angle WZX = \angle YZX$
The included corresponding sides are equal.
$XZ = XZ$
ASA

c. Both triangles are right triangles.
$\angle Q = \angle S = 90°$
The hypotenuses are congruent.
$PR = TR$
A pair of corresponding legs are equal.
$QR = SR$
HL

Similar Triangles

Triangles are similar when all corresponding angles are congruent and when all corresponding sides are in the same ratio.

For a pair of triangles, sets of specific measurements can prove their similarity. The following are three theorems that can prove triangles similar:

AA (angle angle): Two pairs of corresponding angles are equal.

SSS (side side side): All three pairs of corresponding sides are in the same ratio.

SAS (side angle side): Two pairs of corresponding sides are in the same ratio and the included angles are equal.

Example:

Prove $\triangle ABC \sim \triangle ADE$.

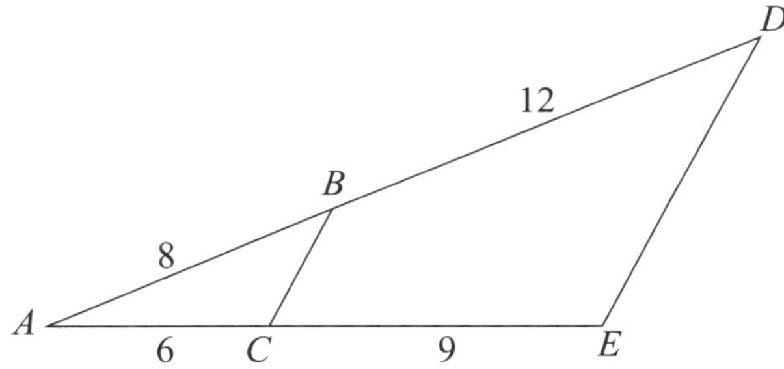

$\overline{AB}$ corresponds to $\overline{AD}$

$$\frac{AB}{AD} = \frac{8}{20} = \frac{2}{5}$$

$\overline{AC}$ corresponds to $\overline{AE}$

$$\frac{AC}{AE} = \frac{6}{15} = \frac{2}{5}$$

The corresponding sides are in the same ratio of $\frac{2}{5}$. The included angle for both triangles is $\angle A$, so $\angle A \cong \angle A$. Therefore, $\triangle ABC \sim \triangle ADE$ by SAS.

Measurements in Similar Triangles

As previously stated, the corresponding sides of similar triangles are in the same ratio and their corresponding angles are congruent. If the ratio of the corresponding side lengths is $\frac{a}{b}$, the ratio of the areas is $\frac{a^2}{b^2}$.

Example:

Given two similar triangles, the area of the first is 27 in^2. If the lengths of the sides in the first triangle are triple those of the second, what is the area of the second triangle?

The ratio of the side lengths is $\frac{3}{1}$ because the first triangle is triple (three times) the second. Therefore, the ratio of the areas is $\frac{3^2}{1^2} = \frac{9}{1}$. To find the area of the second triangle set up the following proportion:

$$\frac{9}{1} = \frac{27}{x}$$

$$9x = 27$$

$$\frac{9x}{9} = \frac{27}{9}$$ Divide both sides by 9

$$x = 3$$

The area of the second triangle is 3 in^2.

Exercise 2

1. **Determine if $\triangle LMN \cong \triangle WXY$.**

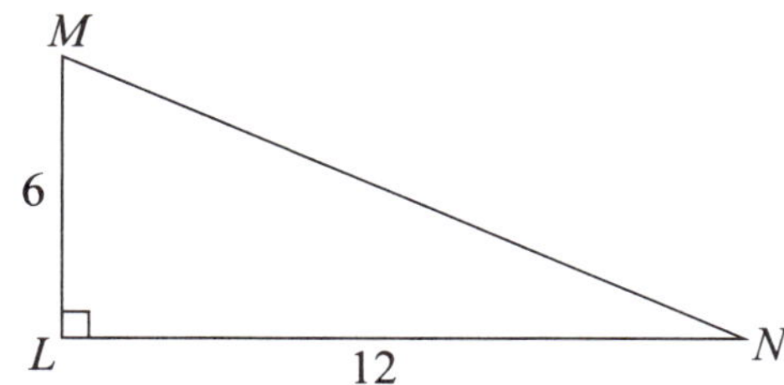

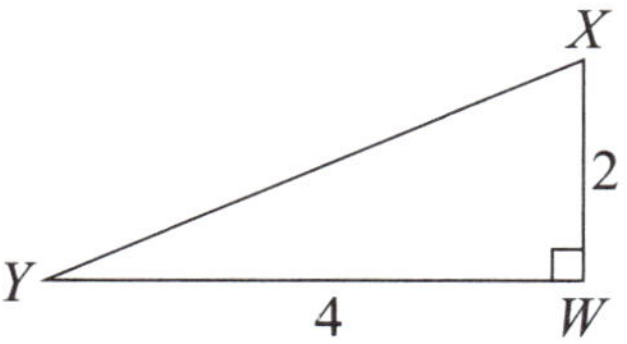

2. **Melissa stated that $\triangle ABC \cong \triangle CDA$ by the AAS theorem. Is she correct?**

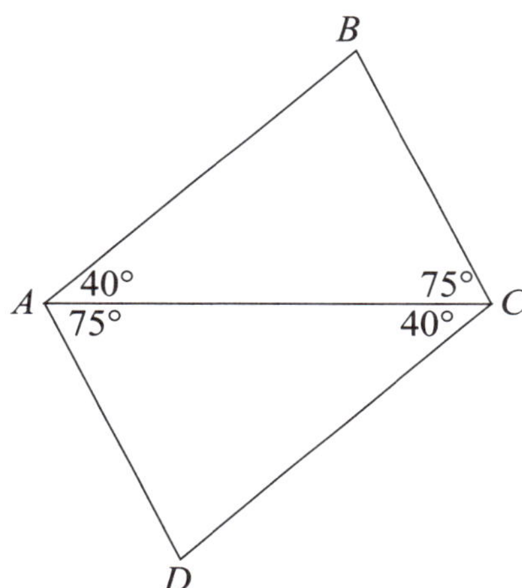

3. **Given two similar triangles, the area of the first triangle is 20 ft^2 and the area of the second triangle is 45 ft^2. What is the ratio of their corresponding side lengths?**

Proportional Segments

If a line is parallel to one side of a triangle and intersects the other two sides, then is divides these sides proportionally.

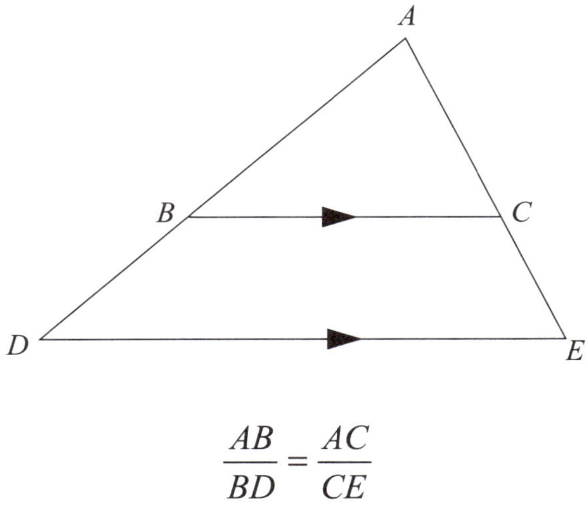

$$\frac{AB}{BD} = \frac{AC}{CE}$$

If a ray bisects one angle of a triangle then it divides the opposite side into segments that are proportional to the two sides that form the bisected angle.

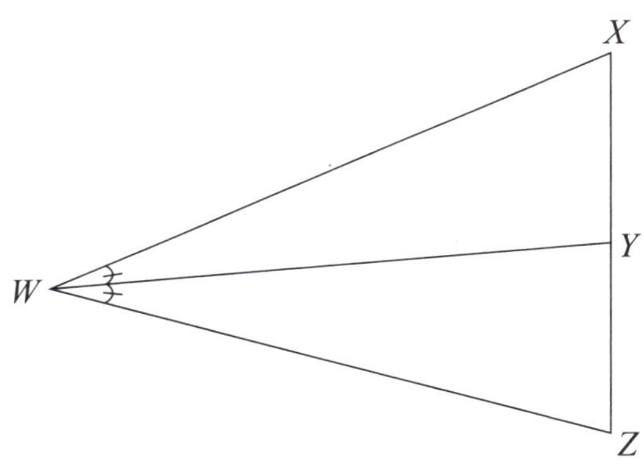

$$\frac{XY}{WX} = \frac{ZY}{WZ} \text{ or } \frac{WX}{WZ} = \frac{XY}{ZY}$$

Example:

Find LN.

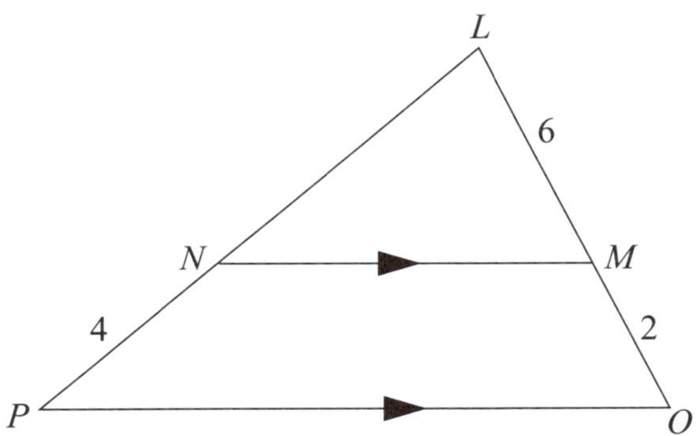

$$\frac{x}{4} = \frac{6}{2}$$

$2x = 24$ Cross multiply

$x = 12$ Divide both sides by 2

$x = 12$

$LN = 12$

Example:

Find ZY.

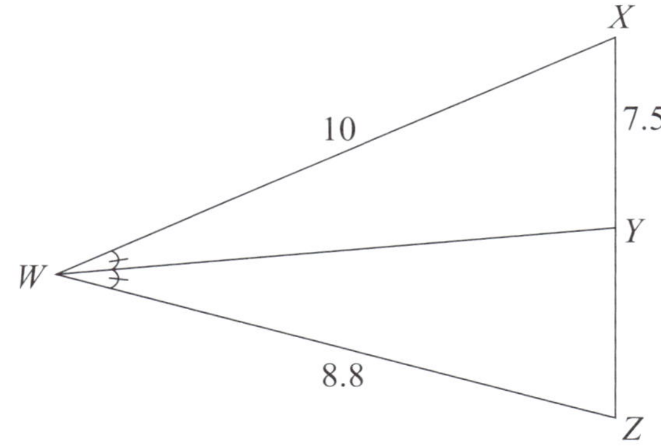

$$\frac{7.5}{10} = \frac{x}{8.8}$$

$10x = 66$ Cross multiply

$$\frac{10x}{10} = \frac{66}{10}$$ Divide both sides by 10

$x = 6.6$

$ZY = 6.6$

Inequality Theorems

Triangle Inequality Theorem:

The sum of the lengths of any two sides of a triangle must be greater than the third side.

4 in

3 in

6 in

$3 + 4 > 6$

$3 + 6 > 4$

$4 + 6 > 3$

Side Angle Inequality Theorems:

In a triangle, the largest angle is across from the longest side.

In a triangle, the smallest angle is across from the shortest side.

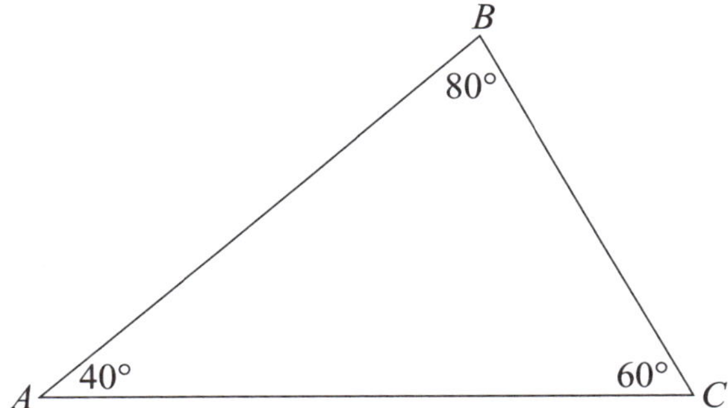

$\overline{AC}$ is the longest side because it is across from the largest angle, $\angle B$.

$\overline{BC}$ is the shortest side because it is across from the smallest angle, $\angle A$.

Exterior Angle Inequality Theorem:

The measure of an exterior angle of a triangle is greater than the measure of either remote interior angle.

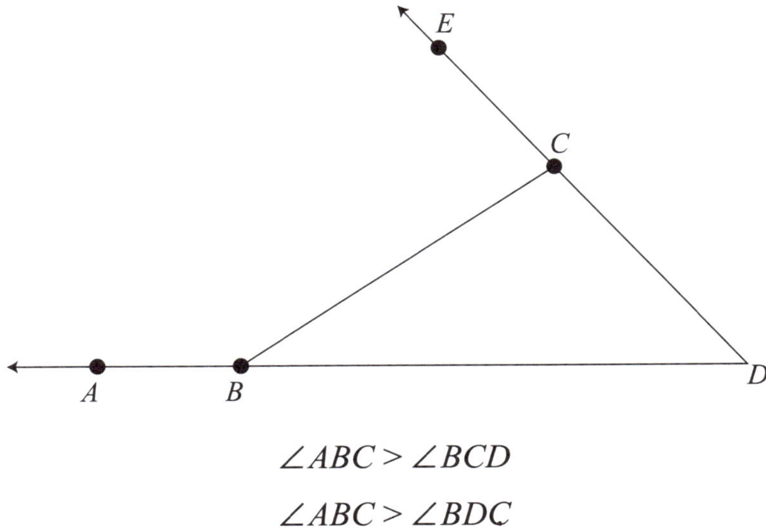

$$\angle ABC > \angle BCD$$

$$\angle ABC > \angle BDC$$

Hinge Theorem:

If two sides of one triangle are congruent to two sides of another triangle and the included angle of the first triangle is larger than that of the second, then the third side of the first triangle is longer than the third side of the second triangle.

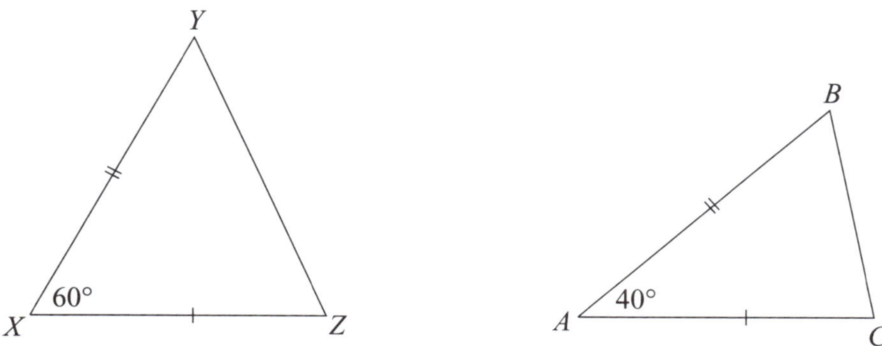

Since $AC = XZ$, $AB = XY$, and $m\angle X > m\angle A$, then $YZ > BC$.

Example:

What are the possible lengths of the third side of the following triangle?

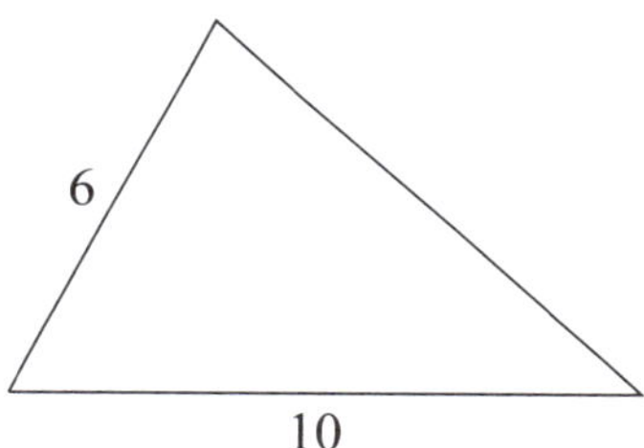

According to the Triangle Inequality Theorem,

$6 + x > 10$	$6 + 10 > x$	$x + 10 > 6$
so, $x > 4$	$x < 16$	$x > -4$ ← *lengths cannot be negative*

The possible values of the missing side are defined by $4 < x < 16$.

Example:

Determine which side of the triangle is the shortest.

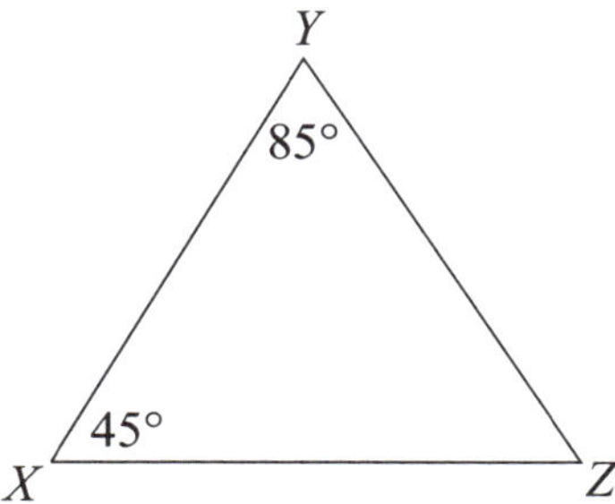

The Side Angle Inequality Theorem states that the shortest side is across from the smallest angle.

Find the missing angle to determine which angle is the smallest.

$45° + 85° + \angle Z = 180°$

$130° + \angle Z = 180°$

$\angle Z = 50°$

Therefore, $\angle X$ is the smallest angle. The side across from $\angle X$ is $\overline{YZ}$.

$\overline{YZ}$ is the shortest side.

Exercise 3

1. Can you draw a triangle with side lengths 3 in., 5 in., and 9 in.?

2. List the sides of $\triangle ABC$ in order from shortest to longest.

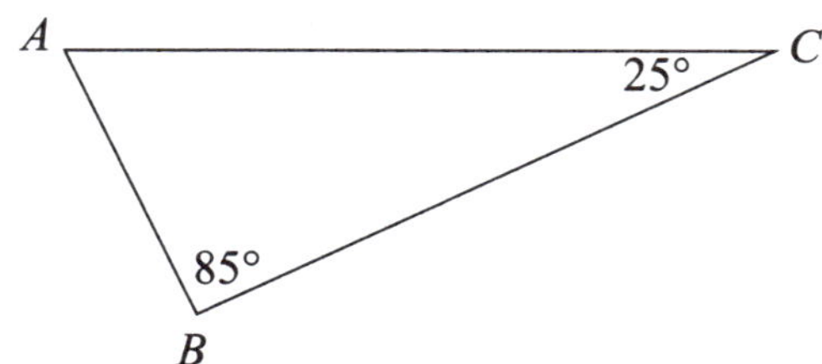

3. Find *RT*.

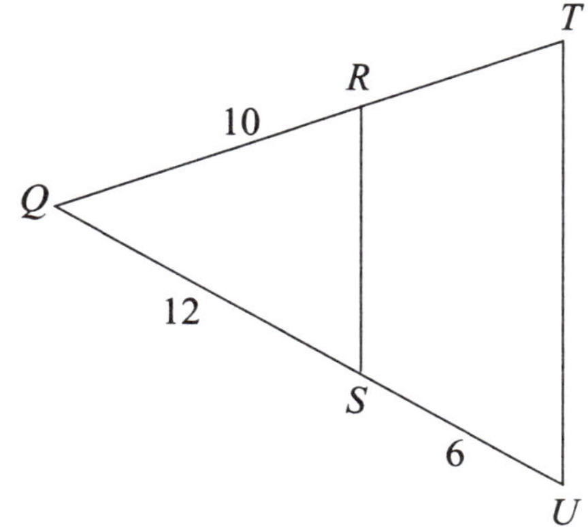

End-of-Chapter Quiz

1. The circumcenter of a triangle ______________ lies outside the triangle.

A. always

B. sometimes

C. never

D. cannot be determined

2. **Which of the following pairs of triangles can be proved congruent by the SAS postulate?**

A.

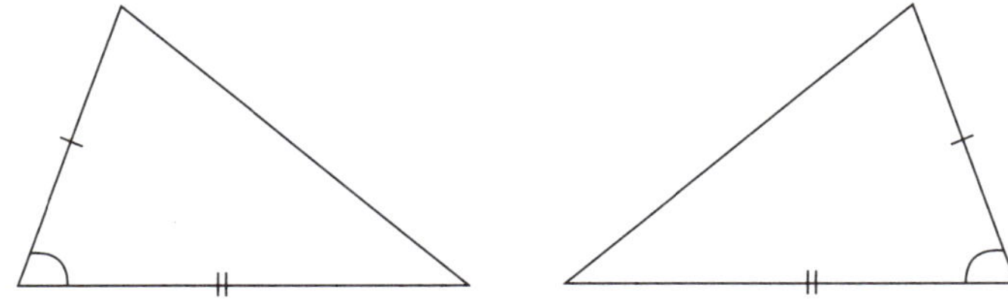

B.

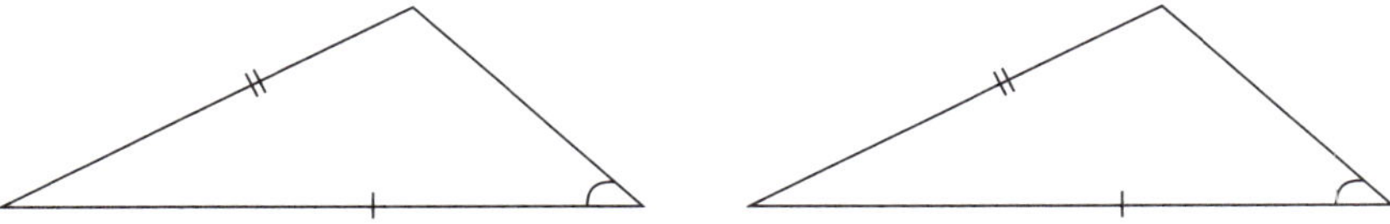

C. both A and B

D. neither A or B

3. **Which of the following can be the length of side *AC*?**

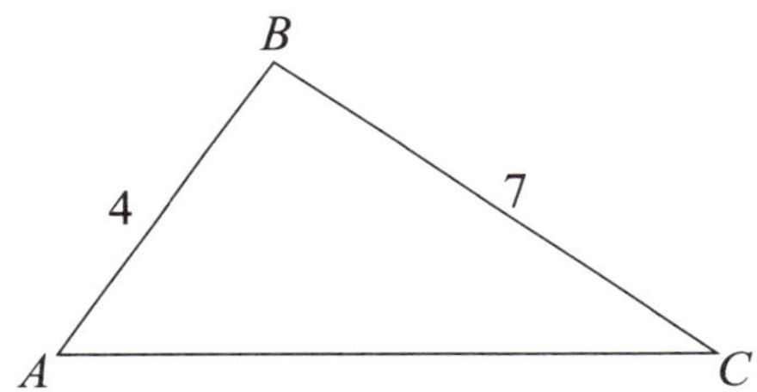

A. 10

B. 11

C. 12

D. 13

4. Which side of $\triangle EFG$ is the shortest given $m\angle E = 40°$, $m\angle F = 60°$, and $m\angle G = 80°$?

A. EF

B. EG

C. FG

D. They are all equal.

5. Given that $AB = LM$ and $AC = LN$, which of the following can be concluded:

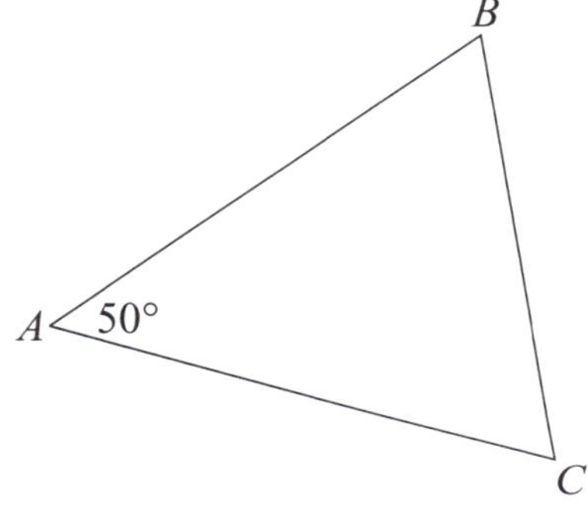

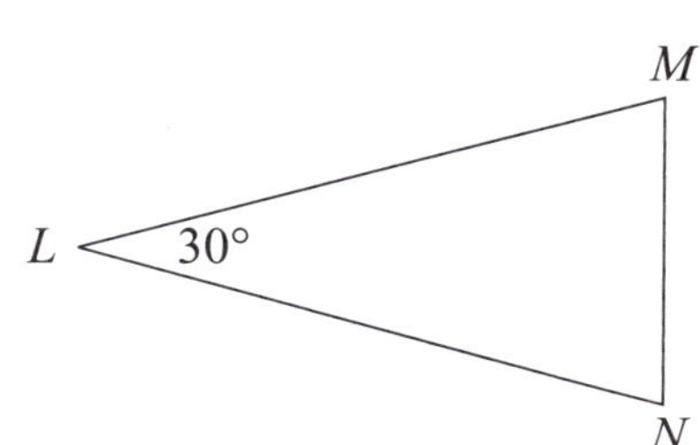

A. $BC = MN$

B. $BC > MN$

C. $BC < MN$

D. not enough information

6. Which one of the following sets of three lengths CANNOT form the sides of a triangle?

A. 8, 9, 12

B. 4, 6, 11

C. 3, 5, 6

D. 10, 12, 20

7. **Look at the following triangles, with their markings.**

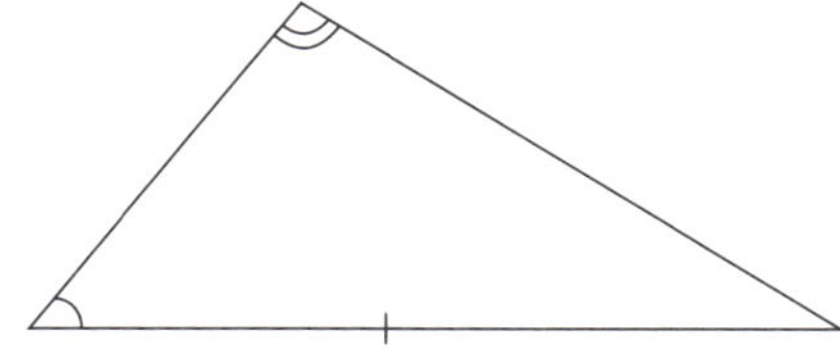
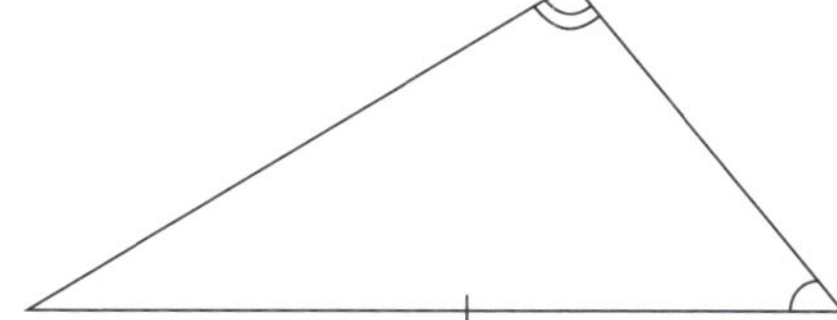

These triangles are shown to be congruent by which of the following?

A. Side-Side-Side

B. Side-Angle-Side

C. Angle-Side-Angle

D. Angle-Angle-Side

8. **In triangle *ABC*, $m\angle A = 55°$ and $m\angle B = 45°$. Which side is the longest?**

A. $\overline{AB}$

B. $\overline{AC}$

C. $\overline{BC}$

D. cannot be determined

9. **Point *P* is equidistant from the three vertices of $\triangle XYZ$. This means that *P* is the ___________ of $\triangle XYZ$.**

A. centroid

B. circumcenter

C. orthocenter

D. incenter

For questions 10–13, use the figure below.

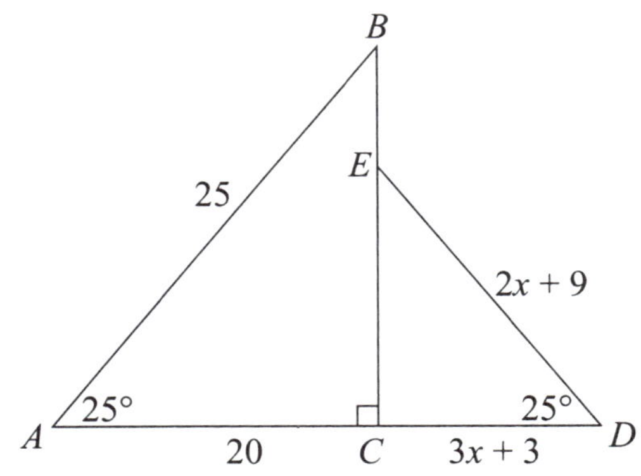

10. Which postulate is used to prove the two triangles are similar?

A. AA Similarity

B. SSS Similarity

C. SAS Similarity

D. The figures are not similar

11. Which of the following is the $m\angle DEC$.

A. 25°

B. 65°

C. 90°

D. 115°

12. Which of the following is the length of $\overline{CD}$?

A. 3

B. 12

C. 15

D. 20

13. If $BC = 15$, which of the following is the length of EC?

A. 3

B. 9

C. 12

D. 20

14. P is the centroid of $\triangle WXY$.

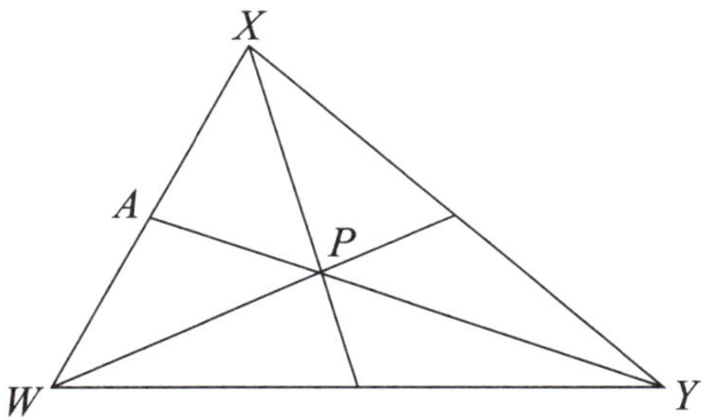

If $PY = 8$, what is the value of AP?

A. 4

B. 8

C. 12

D. 20

15. A triangle has an area of 144 square inches. If a similar triangle has an area twice as large, how does each side length change?

A. The side lengths are multiplied by 2.

B. The side lengths are multiplied by 4.

C. The side lengths are multiplied by $\sqrt{2}$.

D. The side lengths are multiplied by $\sqrt{3}$.

Answers to Exercise 1

1. All 3 angles of a triangle cannot be obtuse because the sum of the angles would be greater than 180°.

2. 12

$$XY = \frac{2}{3}XZ$$

$$8 = \frac{2}{3}XZ$$

$$12 = XZ$$

3. (0, –2)

Find the slope of $\overline{AB}$ and $\overline{BC}$

AB	BC
$m = \dfrac{y_2 - y_1}{x_2 - x_1}$	$m = \dfrac{y_2 - y_1}{x_2 - x_1}$
$m = \dfrac{-2 - 2}{-4 - 0}$	$m = \dfrac{-2 - 2}{4 - 0}$
$m = \dfrac{-4}{-4}$	$m = \dfrac{-4}{4}$
$m = 1$	$m = -1$

Find the midpoint of $\overline{AB}$ and $\overline{BC}$

AB	BC
$M = \left(\dfrac{x_1 + x_2}{2}, \dfrac{y_1 + y_2}{2}\right)$	$M = \left(\dfrac{x_1 + x_2}{2}, \dfrac{y_1 + y_2}{2}\right)$
$M = \left(\dfrac{-4 + 0}{2}, \dfrac{-2 + 2}{2}\right)$	$M = \left(\dfrac{0 + 4}{2}, \dfrac{2 + -2}{2}\right)$
$M = (-2,\ 0)$	$M = (2,\ 0)$

Find the perpendiculor bisector of $\overline{AB}$ and $\overline{BC}$

If the slope of AB is 1, the perpendicular bisector has a slope of –1 and passes through the midpoint (–2, 0).

$y = mx + b$

$0 = -1(-2) + b$

$0 = 2 + b$ ⟶ $y = -x - 2$ perpendicular bisector of $\overline{AB}$

$-2 = b$

If the slope of BC is –1, the perpendicular bisector has a slope of 1 and passes through the midpoint (2, 0).

$y = mx + b$

$0 = 1(2) + b$

$0 = 2 + b$ ⟶ $y = x - 2$ perpendicular bisector of $\overline{BC}$

$-2 = b$

The intersection of the perpendicular bisectors is the location of the circumcenter.

$-x - 2 = x - 2$

$2x = 0$

$x = 0$

Substitute $x = 0$ into $y = x - 2 \Rightarrow y = 0 - 2 \Rightarrow y = -2$

The circumcenter is located at (0, –2).

Answers to Exercise 2

1. Yes

$$\frac{ML}{XW} = \frac{6}{2} = 3$$

$$\frac{LN}{WY} = \frac{12}{4} = 3$$

$\angle MLN \cong \angle XWY$

Two pairs of corresponging sides are in the same ratio and the included angles are equal. $\triangle MLN \cong \triangle XWY$ by the SAS Similarity Postulate.

2. No

$m\angle BAC = m\angle DCA$

$m\angle BCA = m\angle DAC$

$AC = AC$

Since AC is the included side between the congruent angles, Melissa is incorrect. $\triangle ABC \cong \triangle CDA$ by the ASA Postulate

3. $\frac{2}{3}$

$$\frac{20}{45} = \frac{4}{9}$$

$$\frac{a^2}{b^2} = \frac{4}{9}$$

$$\frac{a}{b} = \frac{\sqrt{4}}{\sqrt{9}} = \frac{2}{3} \text{ ratio of side lengths}$$

Answers to Exercise 3

1. No

The sum of 3 and 5 is not greater than 9, therefore, the lengths 3 in, 5 in, and 9 in do not result in a triangle.

2. $m\angle A = 70°$

$\overline{AB}, \overline{BC}, \overline{AC}$

3. 5

$$\frac{12}{18} = \frac{10}{10+x}$$

$$120 + 12x = 180$$

$$12x = 60$$

$$x = 5$$

$$RT = 5$$

OR

$$\frac{10}{x} = \frac{12}{6}$$

$$12x = 60$$

$$x = 5$$

$$RT = 5$$

Answers to Chapter Quiz

1. B

The circumcenter can be inside, on, or outside the triangle.

2. A

Two pairs of corresponding sides and their included angles are congruent.

3. A

$4 + 7 > 10$

4. C

$\overline{FG}$ is the shortest side because it is across from the smallest angle, $\angle E$.

5. B

According to the Hinge Theorem, $BC > MN$ because the angle $m\angle BAC > m\angle MLN$.

6. B

The sum of 4 and 6 is not greater than 11.

7. D

Two corresponding angles are congruent and a pair of non-included corresponding sides are congruent.

8. A

$m\angle C = 80°$, so the longest side is $\overline{AB}$.

9. B

Since P is equidistant from all three vertices, it must lie on the perpendicular bisector of each side. By definition, the circumcenter lies at the intersection of all three perpendicular bisectors.

10. A

$\triangle ABC$ and $\triangle DEC$ have two congruent angles.

$m\angle BAC = m\angle EDC$

$m\angle BCA = m\angle ECD$

11. B

$m\angle ECD + m\angle EDC + m\angle DEC = 180°$

$90° + 25° + m\angle DEC = 180°$

$115° + m\angle DEC = 180°$

$m\angle DEC = 65°$

12. B

$$\frac{20}{3x+3} = \frac{25}{2x+9}$$

$40x + 180 = 75x + 75$

$105 = 35x$

$3 = x$

Substitute $x = 3$ into $3x + 3 \Rightarrow 3(3) + 3 = 12$

13. B

In question 12, it was determined that $CD = 12$. Therefore, $\frac{20}{12} = \frac{15}{x} \Rightarrow 20x = 180 \Rightarrow x = 9$.

14. A

$PY = \frac{2}{3}AY$

$8 = \frac{2}{3}AY$

$12 = AY$

$AY = PY + AP$

$12 = 8 + AP$

$4 = AP$

15. C

If the area is doubled, the ratio of the areas is $\frac{144}{288} = \frac{1}{2}$. In similar triangles, the ratio of the corresponding side lengths is $\frac{a}{b}$ and the ratio of the areas is $\frac{a^2}{b^2}$. Therefore,

$\frac{a^2}{b^2} = \frac{1}{2}$

$\frac{a}{b} = \frac{\sqrt{1}}{\sqrt{2}} = \frac{1}{\sqrt{2}}$

The side lengths are in a ratio of $1 : \sqrt{2}$.

Chapter 5

Right Triangles

Your Goals for Chapter 5

1. You should be able to apply the Pythagorean Theorem to solve problems.
2. You should be able to use the properties of special right triangles to solve problems.
3. You should be able to apply the relationships created when an altitude is drawn to a hypotenuse.
4. You should be able to use trigonometric ratios to solve problems.

Standards

The following standards are assessed on Florida's Geometry End-of-Course exam either directly or indirectly:

MA.912.G.5.1: (High) Prove and apply the Pythagorean Theorem and its converse.

MA.912.G.5.2: (Moderate) State and apply the relationships that exist when the altitude is drawn to the hypotenuse of a right triangle.

MA.912.G.5.3: (Moderate) Use special right triangles (30°–60°–90° and 45°–45°–90°) to solve problems.

MA.912.G.5.4: (High) Solve real-world problems involving right triangles.

MA.912.T.2.1: (Moderate) Define and use the trigonometric ratios (sine, cosine, tangent, cotangent, secant, cosecant) in terms of angles of right triangles.

Pythagorean Theorem

If a triangle is a right triangle and a square is made from each of the three sides, then the biggest square has the same area as the sum of the areas of the other two squares.

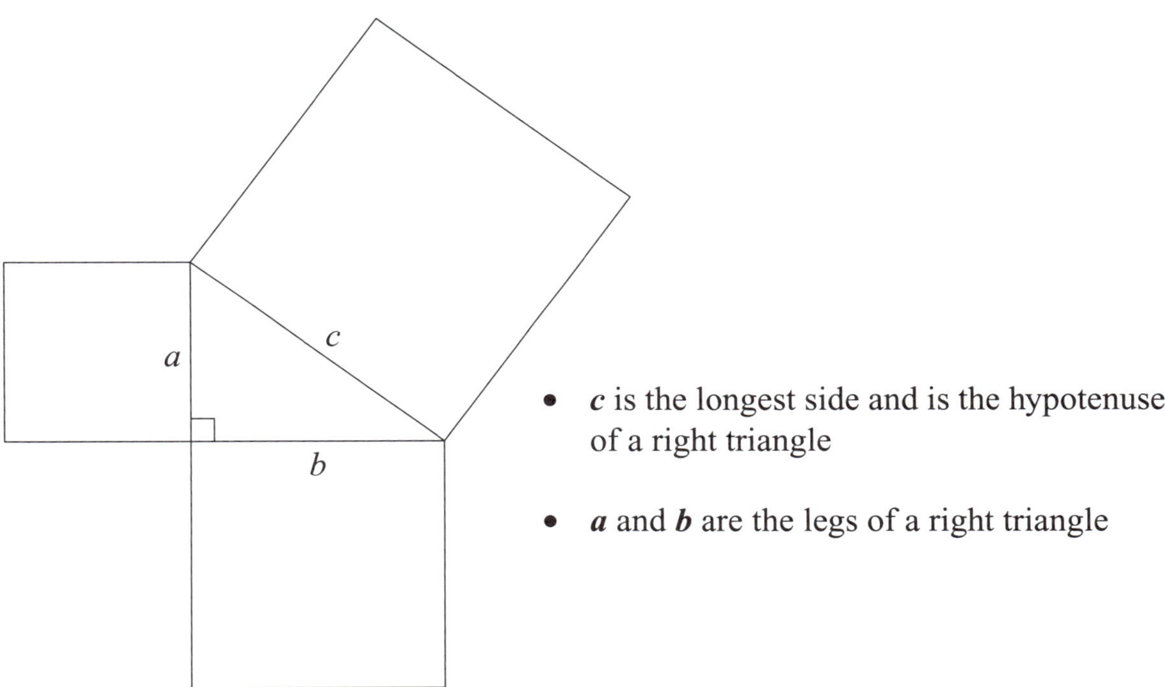

- ***c*** is the longest side and is the hypotenuse of a right triangle
- ***a*** and ***b*** are the legs of a right triangle

Given a right triangle, the square of the hypotenuse is equal to the sum of the squares of the other 2 sides. This is the **Pythagorean Theorem**.

$$a^2 + b^2 = c^2$$

Note: c is always the hypotenuse, the side across from the right angle.

Example:

Find the length of the missing side.

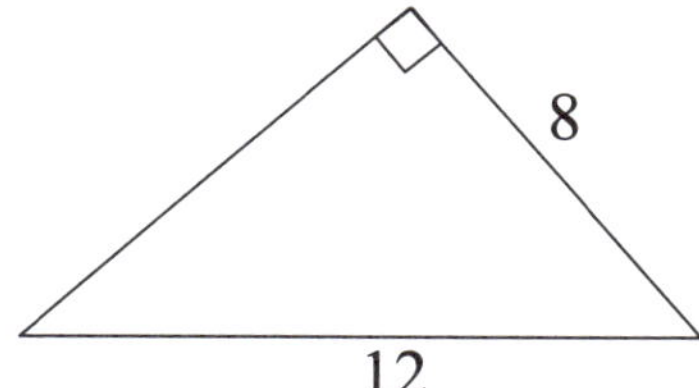

Since the triangle is a right triangle, the Pythagorean Theorem can be used to determine the length of the missing side.

$a^2 + b^2 = c^2$

$a^2 + 8^2 = 12^2$

$a^2 + 64 = 144$

$a^2 + 64 - 64 = 144 - 64$

$a^2 = 80$

$\sqrt{a^2} = \sqrt{80}$

$a = \sqrt{16 \cdot 5}$

$a = 4\sqrt{5}$

Example:

Jake lives at point B. To reach his friend's house at point D, he walks 34 meters to point A and 41 meters to point D. How many meters would be saved if it were possible to walk a straight line from point B to point D?

A straight line drawn directly from point B to point D would result in two right triangles. The line drawn would be the hypotenuse.

$a^2 + b^2 = c^2$

$34^2 + 41^2 = c^2$

$1156 + 1681 = c^2$

$2837 = c^2$

$\sqrt{2837} = \sqrt{c^2}$

$53.26 \approx c$

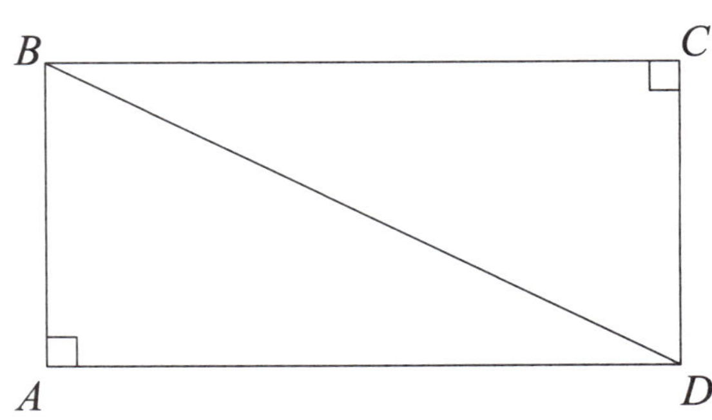

The walk directly from point B to point D would be approximately 53.26 meters. Jake's walk from point B to point A to point D is 34 meters + 41 meters = 75 meters. By walking directly from point B to point D, Jake would save approximately 21.74 meters (75 meters – 53.26 meters).

The Pythagorean Theorem can be used to determine if a triangle is acute, right, or obtuse.

If $c^2 < a^2 + b^2$, then the triangle is acute.

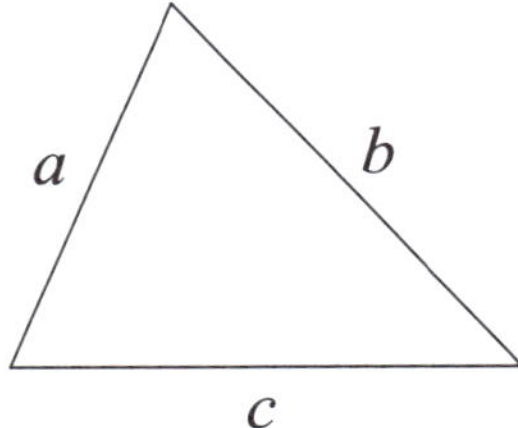

If $c^2 = a^2 + b^2$, then the triangle is right.

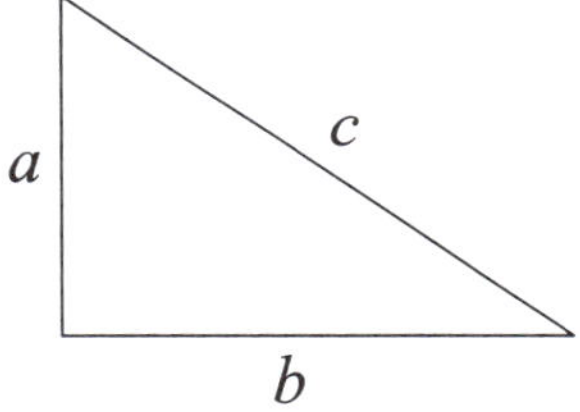

If $c^2 > a^2 + b^2$, then the triangle is obtuse.

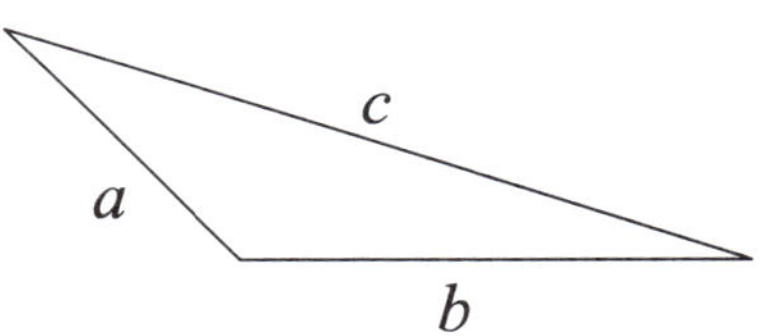

Example:

Determine if a triangle with side lengths 6, 7, and 8 is acute, right, or obtuse.

Since the hypotenuse is the longest side, assume that c = 8.

8^2 ___ $6^2 + 7^2$

64 ___ $36 + 49$

$64 < 85$

The triangle is acute.

Altitude to the Hypotenuse

In right triangle *WXY*, the altitude $\overline{WZ}$ is drawn to the hypotenuse $\overline{XY}$.

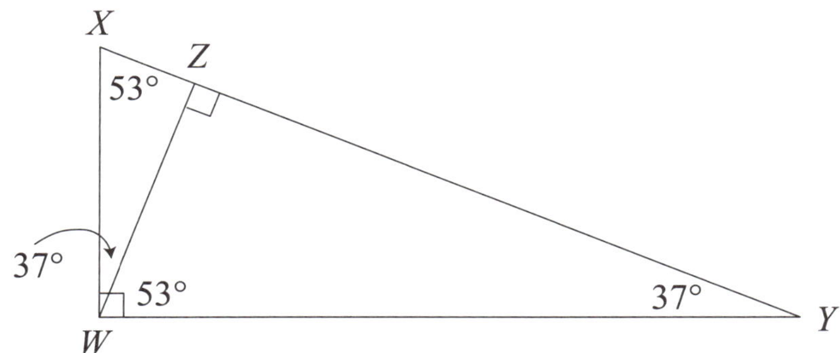

The altitude drawn creates three similar triangles. The two smaller triangles created are similar to each other and similar to the original triangle. Thus, $\triangle WZX \sim \triangle YZW \sim \triangle YWX$

Because the triangles are similar, the following theorems can be stated:

1) If an altitude is drawn to the hypotenuse of a right triangle, then each leg is the geometric mean between the hypotenuse and its touching segment on the hypotenuse.

$$\frac{XY}{XW} = \frac{XW}{XZ}$$

and

$$\frac{YX}{YW} = \frac{YW}{YZ}$$

2) If an altitude is drawn to the hypotenuse of a right triangle, then it is the geometric mean between the segments of the hypotenuse.

$$\frac{XZ}{WZ} = \frac{WZ}{YZ}$$

Example:

Find the value of x and y.

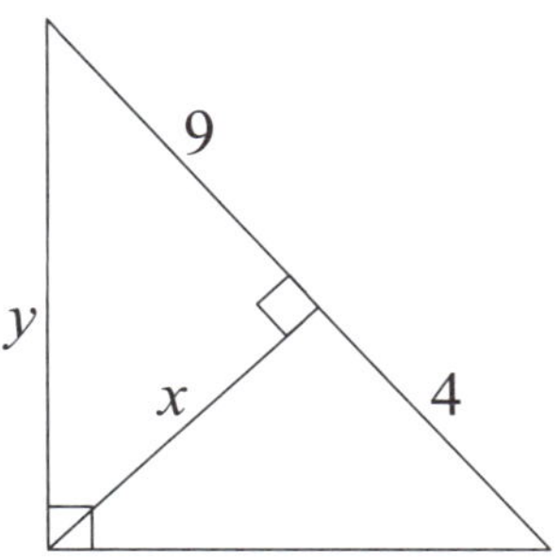

To find the value of x,

$$\frac{9}{x} = \frac{x}{4}$$

$$36 = x^2$$

$$\sqrt{36} = \sqrt{x^2}$$

$$6 = x$$

To find the value of y,

$$\frac{13}{y} = \frac{y}{9}$$

$$117 = y^2$$

$$\sqrt{117} = \sqrt{y^2}$$

$$117 = y$$

$$10.82 \approx y$$

Note: You will notice that y can be found using the Pythagorean Theorem: $6^2 + 9^2 = y^2$.

Special Right Triangles

A **30°-60°-90° triangle** is a type of special right triangle whose angles measure 30°, 60°, and 90°.

The lengths of the sides of a 30°-60°-90° triangle are in a ratio of $1:\sqrt{3}:2$.

30°-60°-90° Triangle

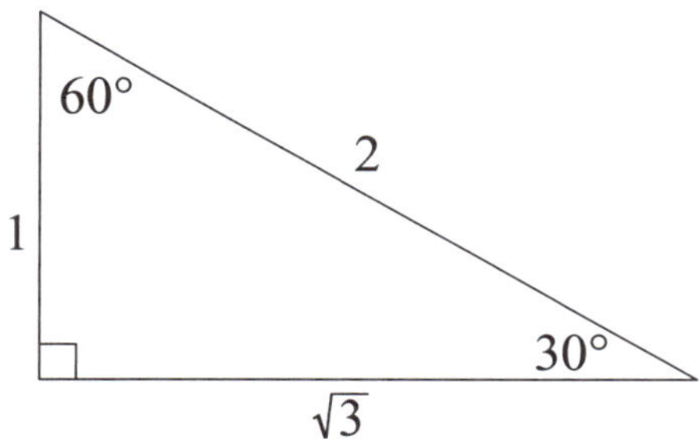

Note: The shortest side is across from the smallest angle and the longest side is across from the largest angle.

Note: The altitude of an equilateral triangle forms two 30°-60°-90° triangles.

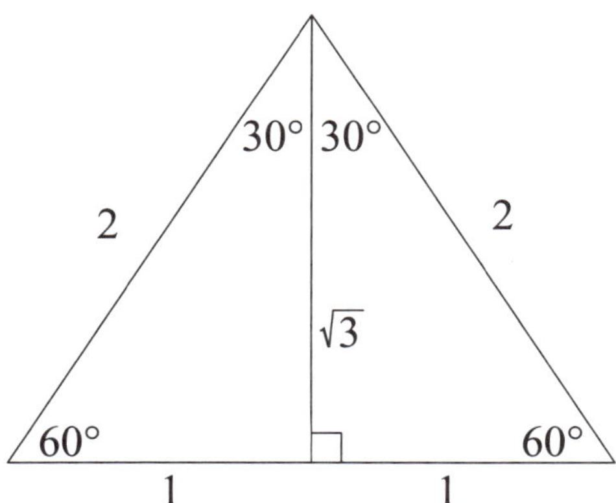

A **45°-45°-90° triangle** is a type of special right triangle whose angles measure 45°, 45°, and 90°.

The lengths of the sides of a 45°-45°-90° triangle are in a ratio of $1:1:\sqrt{2}$.

45°-45°-90° Triangle

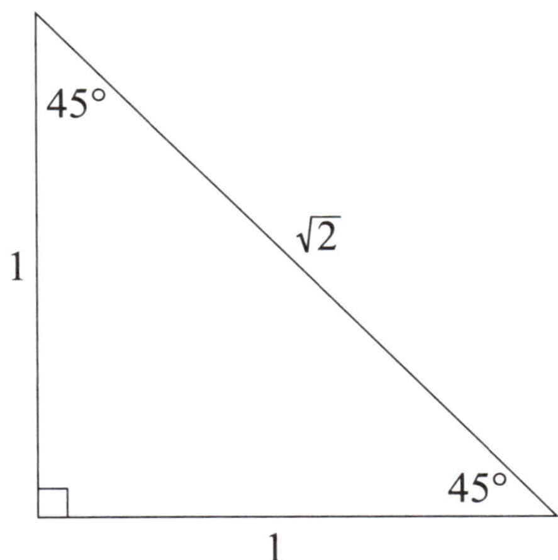

Note: A 45°-45°-90° triangle has two sides of equal length; therefore, it is an isosceles right triangle.

Example:

Determine the value of x and y.

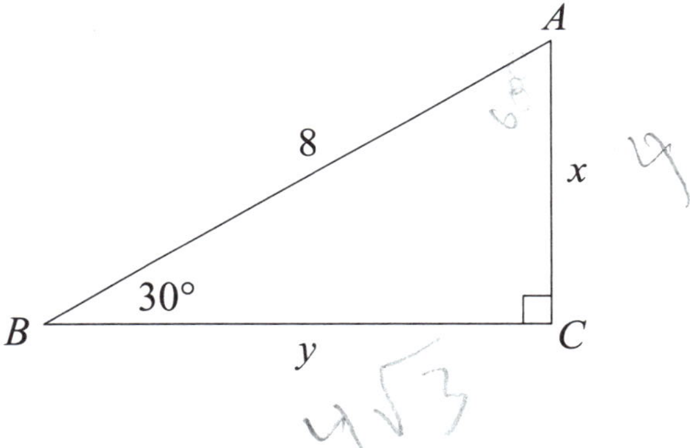

Since the hypotenuse is 8, the side opposite the 30° angle must be $\left(\frac{1}{2}\right)(8) = 4$. The length of the side opposite the 60° angle, $\angle A$, is found by multiplying the length of the shortest side by $\sqrt{3}$. Thus, the length of the side opposite the 60° angle is $4\sqrt{3}$. Therefore, $x = 4$ and $y = 4\sqrt{3}$.

Example:

Determine the value of x and y.

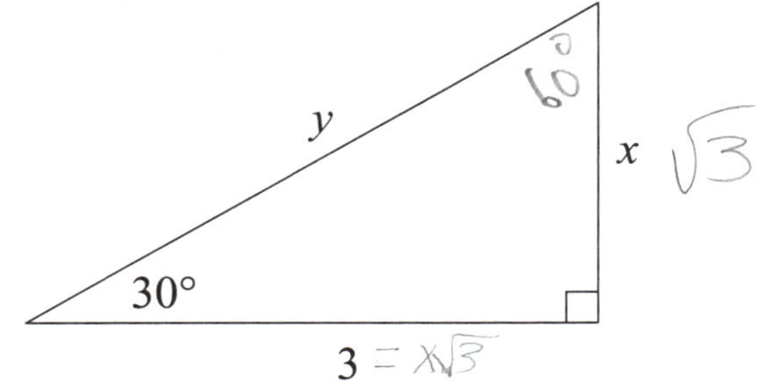

Using the 30°-60°-90° triangle from above,

$a = x$ $\quad\quad$ $a\sqrt{3} = 3$ $\quad\quad$ $2a = y$

Solve: $a\sqrt{3} = 3$

$$\frac{a\sqrt{3}}{\sqrt{3}} = \frac{3}{\sqrt{3}}$$

$$a = \frac{3}{\sqrt{3}} \cdot \frac{\sqrt{3}}{\sqrt{3}}$$

$$a = \frac{3\sqrt{3}}{3} = \sqrt{3}$$

Since $a = \sqrt{3}$,

$a = x$ $\quad\quad$ $2a = y$

$\sqrt{3} = x$ $\quad\quad$ $2\sqrt{3} = y$

The length of side x is $\sqrt{3}$ and the length of side y is $2\sqrt{3}$.

Example:

Determine the value of x.

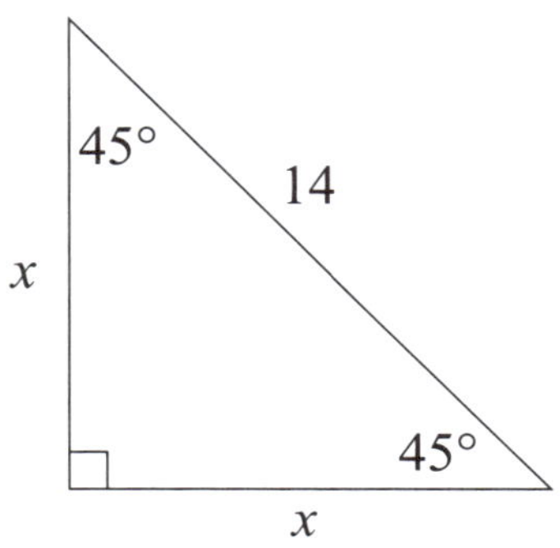

Using the 45°-45°-90° triangle from above,

$$x\sqrt{2} = 14$$

Solve: $x\sqrt{2} = 14$

$$\frac{x\sqrt{2}}{\sqrt{2}} = \frac{14}{\sqrt{2}}$$

$$x = \frac{14}{\sqrt{2}} \cdot \frac{\sqrt{2}}{\sqrt{2}}$$

$$x = \frac{14\sqrt{2}}{2} = 7\sqrt{2}$$

The length of side x is $7\sqrt{2}$.

Example:

Given a right triangle with the two legs each measuring 5 inches, determine the length of the hypotenuse.

Since this is a right triangle with 2 equal sides, it is a 45°-45°-90° triangle.

The legs measure 5 inches, therefore $a = 5$.

Since $a = 5$, $a\sqrt{2} = 5\sqrt{2}$.

The length of the hypotenuse is $5\sqrt{2}$.

Trigonometric Ratios

Consider the following right triangle *ABC*.

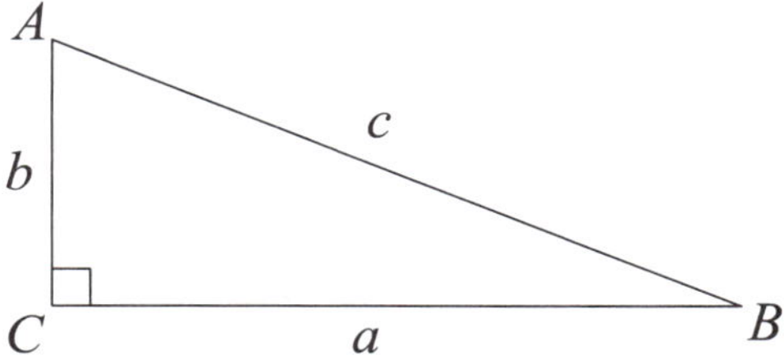

Then the sine of angle A, abbreviated as $\sin \angle A$, equals the ratio of the opposite leg to the hypotenuse. Using the above triangle, this can be expressed as $\sin \angle A = \frac{a}{c}$.

Then the cosine of angle A, abbreviated as $\cos \angle A$, equals the ratio of the adjacent leg to the hypotenuse. Using the above triangle, this can be expressed as $\cos \angle A = \frac{b}{c}$.

Then the tangent of angle A, abbreviated as $\tan \angle A$, equals the ratio of the opposite leg to the adjacent leg. Using the above triangle, this can be expressed as $\tan \angle A = \frac{a}{b}$.

For each of these three trigonometric functions there is a reciprocal function. The cosecant of angle A, abbreviated as $\csc \angle A$, equals the ratio of the hypotenuse to the opposite leg. Using the above triangle, $\csc \angle A = \frac{c}{a}$. Note that $\csc \angle A = \frac{1}{\sin \angle A}$.

The secant of angle A, abbreviated as $\sec A$, equals the ratio of the hypotenuse to the adjacent leg. Using the above triangle, $\sec \angle A = \frac{c}{b}$. Note that $\sec A = \frac{1}{\cos \angle A}$.

The cotangent of angle A, abbreviated as $\cot \angle A$ equals the ratio of the adjacent leg to the opposite leg. Using the above triangle, $\cot \angle A = \frac{b}{a}$. Note that $\cot \angle A = \frac{1}{\tan \angle A}$.

Example:

A telephone pole is supported by a wire extending from the top of the pole to a metal stake in the ground. The wire is 26 feet long and forms a 60° angle with the ground. How tall is the telephone pole?

Let x represent the height of the telephone pole. Draw a picture, as follows:

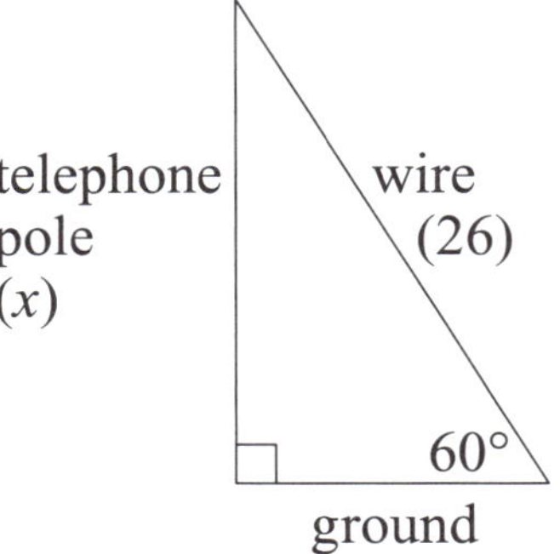

$$\sin 60° = \frac{x}{26}$$

$$.866 = \frac{x}{26}$$

$$22.5 \approx x$$

The building is approximately 22.5 feet tall.

Note: This can also be solved using the 30°-60°-90° triangle ratios.

Example:

Yara is building a 30-foot path diagonally across her square yard as shown below,

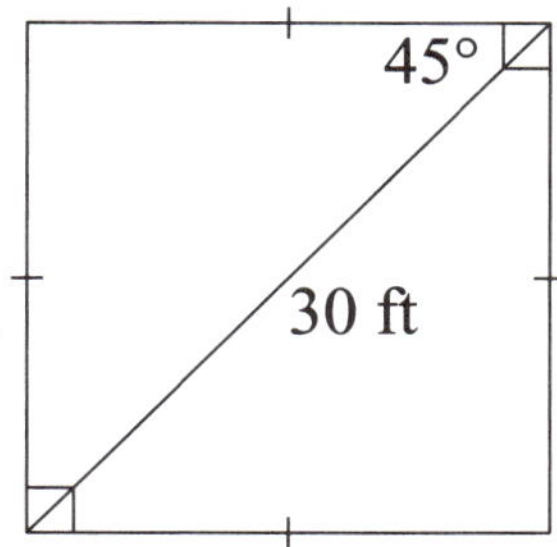

What is the length and width of Yara's yard?

$$\cos 45° = \frac{s}{30}$$

$$.7071 = \frac{s}{30}$$

$$21.2 \approx s$$

Since the yard is a square, the length and width are approximately 21.2 feet.

Note: This can also be solved using the 45°-45°-90° triangle ratios.

End-of-Chapter Quiz

1. The diagonal of a square is 24 inches. What is the length in inches of one side?

A. 12

B. $12\sqrt{2}$

C. $12\sqrt{3}$

D. $24\sqrt{2}$

2. **What are the values of *x* and *y* in the following diagram?**

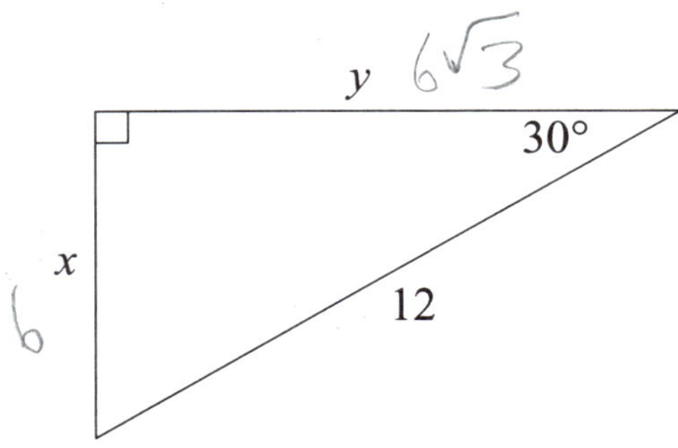

A. $x = 6\sqrt{3}$ and $y = 6$

B. $x = 12$ and $y = 12\sqrt{3}$

C. $x = 6$ and $y = 6\sqrt{3}$

D. $x = 12\sqrt{3}$ and $y = 12$

3. **During a hot air balloon festival you are standing at point *X* and the balloon is located at point *Y*, as shown in the diagram.**

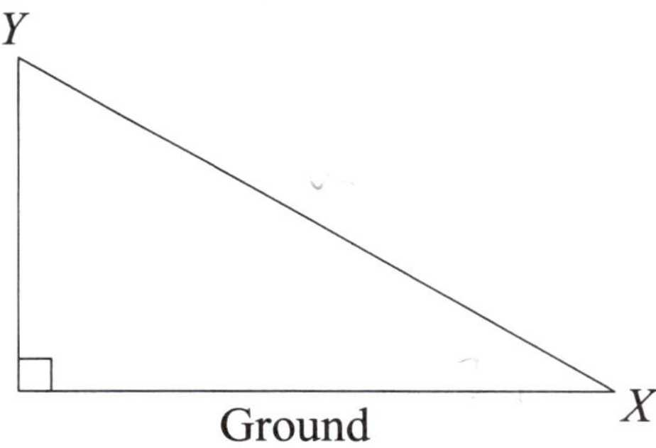

If *XY* = 2 miles and the angle of elevation from *X* is 35°, how many miles up from the ground is the balloon to the nearest hundredth.

A. 0.29

B. 0.41

C. 1.15

D. 1.64

4. The hypotenuse of an isosceles right triangle measures 12 inches, find the length of the legs.

A. $6\sqrt{2}$

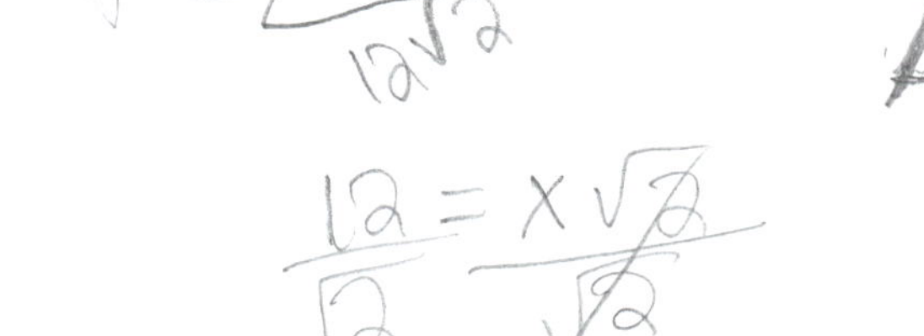

B. 6

C. $12\sqrt{2}$

D. 24

5. Determine if a triangle with side lengths 3, 5, and 7 is acute, right, or obtuse.

6. Lacey wants to measure the height of the tree in her backyard. She stood 30 feet from the base of the tree. The angle of elevation from a point on the ground to the top of the tree is 47°. Estimate the height of the tree.

A. 21.9 feet

B. 32.2 feet

C. 20.5 feet

D. 18.4 feet

For questions 7 and 8, use the right triangle below.

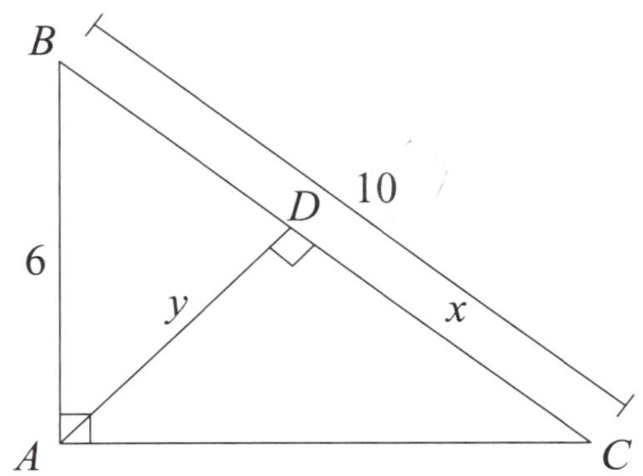

7. Which of the following is the value of *x*?

A. 3.6

B. 4.8

C. 6.4

D. 10

8. Which of the following is the value of *y*?

A. 3.6

B. 4.8

C. 6

D. 6.4

9. Josh is installing a television with a 42-inch diagonal. Their entertainment center is 28 inches high and 32 inches wide. Will the television fit into the entertainment center? Justify your reasoning. Assume that the television and the entertainment center are similar rectangles.

10. A baseball diamond is actually a square. The angle at each base is 90°. If the bases are 90 feet apart, how far is it from home plate to second base?

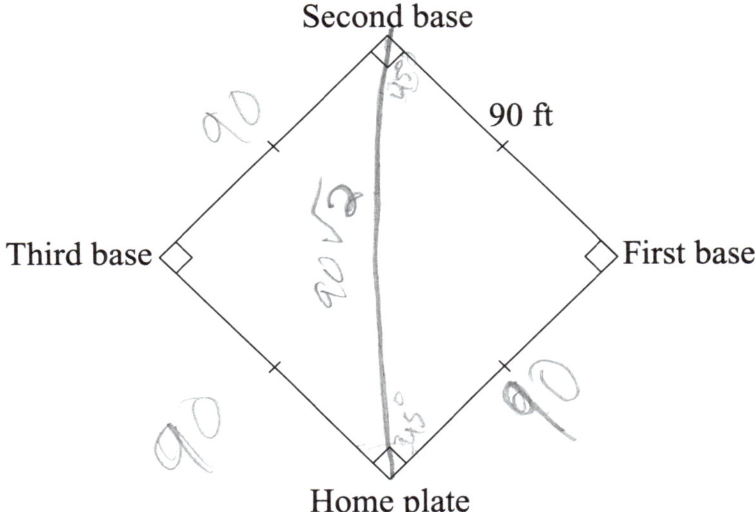

Answers to the Chapter Quiz

1. B

The diagonal of a square represents the hypotenuse of a right triangle in which two adjacent sides are legs. Let s represent the length of any side. Then $s^2 + s^2 = 24^2$, which becomes $2s^2 = 576 \Rightarrow s^2 = 288 \Rightarrow s = \sqrt{288} = \sqrt{144} \times \sqrt{2} = 12\sqrt{2}$.

2. C

Since the hyypotenuse is 12, the side opposite the 30° angle is $\frac{1}{2}(12) = 6$. Therefore, $x = 6$. The remaining leg equals the side opposite 30° times $\sqrt{3}$. Therefore, $y = 6\sqrt{3}$.

3. C

Let h represent the vertical height from the balloon to the ground. Then $\sin 35° = \frac{h}{2}$. Multiply both sides by 2 to get $h = 2(\sin 35°) \approx (2)(0.574) \approx 1.15$.

4. A

An isosceles right triangle is a 45°-45°-90° triangle. Since the hypotenuse is 12, the length of each leg is $\frac{12}{\sqrt{2}}$. This means that $\frac{12}{\sqrt{2}} \cdot \frac{\sqrt{2}}{\sqrt{2}} = \frac{12\sqrt{2}}{2} = 6\sqrt{2}$ is the length of each leg.

5. Obtuse

7^2 ____ $3^2 + 5^2$

49 ____ 9 + 25

$49 > 34$

The triangle is an obtuse triangle.

6. B

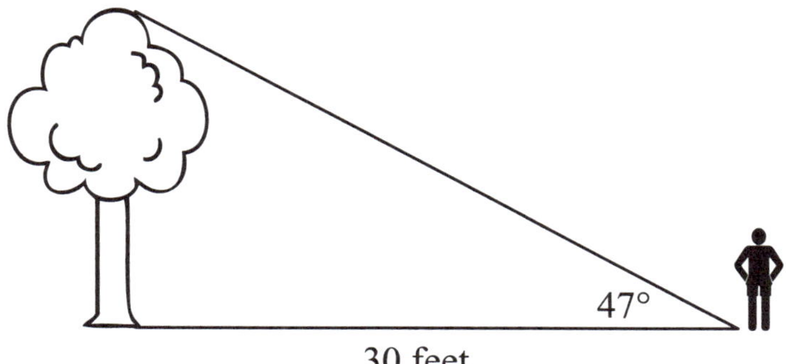

$$\tan 47° = \frac{x}{30}$$

$$30(\tan 47°) = \left(\frac{x}{30}\right)30$$

$$30(1.072) \approx x$$

$$32.2 \approx x$$

7. C

$$\frac{10}{6} = \frac{6}{10 - x}$$

$$\begin{array}{r} 100 - 10x = 36 \\ \underline{-100 \qquad\quad -100} \end{array}$$

$$\frac{-10x}{-10} = \frac{-64}{-10}$$

$$x = 6.4$$

8. B

Use $x = 6.4$ from question 7, so that $BD = 10 - x = 3.6$.

$$\frac{3.6}{y} = \frac{y}{6.4}$$

$$y^2 = 23.04$$

$$y = \sqrt{23.04} = 4.8$$

9. $28^2 + 32^2 = c^2$

$784 + 1024 = c^2$

$1808 = c^2$

$\sqrt{1808} = \sqrt{c^2}$

$42.5 \approx c$

Since the diagonal of the entertainment center is 42.5 and is greater than the 42-inch diagonal of the television, the television will fit.

10. The distance from home plate to second base is the hypotenuse of an isosceles right triangle with legs of 90 ft.

$90^2 + 90^2 = c^2$

$8100 + 8100 = c^2$

$16200 = c^2$

$\sqrt{16200} = \sqrt{c^2}$

$127.3 \approx c$

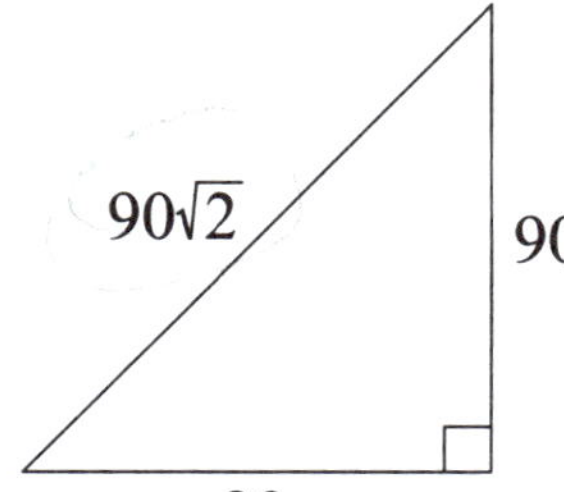

Chapter 6

Polyhedrons and Other Solids

Your Goals for Chapter 6

1. You should be able to classify solid figures and identify their nets.
2. You should be able to determine the number of faces, edges, and vertices in a polyhedron.
3. You should be able to identify chords, tangents, radii, and great circles of spheres.
4. You should be able to find the surface area and volume of solid figures.
5. You should be able to apply properties of congruent and similar polygons and determine the effect on surface area and volume based on changes to dimensions.

Standards

The following standards are assessed on Florida's Geometry End-of-Course exam either directly or indirectly:

MA.912.G.7.1: (Moderate) Describe and make regular, non-regular, and oblique polyhedra, and sketch the net for a given polyhedron and vice versa.

MA.912.G.7.2: (Moderate) Describe the relationships between the faces, edges, and vertices of polyhedra.

MA.912.G.7.4: (Low) Identify chords, tangents, radii, and great circles of spheres

MA.912.G.7.5: (Moderate) Explain and use formulas for lateral area, surface area, and volume of solids.

MA.912.G.7.6: (Moderate) Identify and use properties of congruent and similar solids.

MA.912.G.7.7: (Moderate) Determine how changes in dimensions affect the surface area and volume of common geometric solids.

Solid Figures

A three-dimensional figure can be referred to as a **solid figure**. The most common types of solid figures are shown in the table below.

Prism	Pyramid	Cylinder	Cone	Sphere
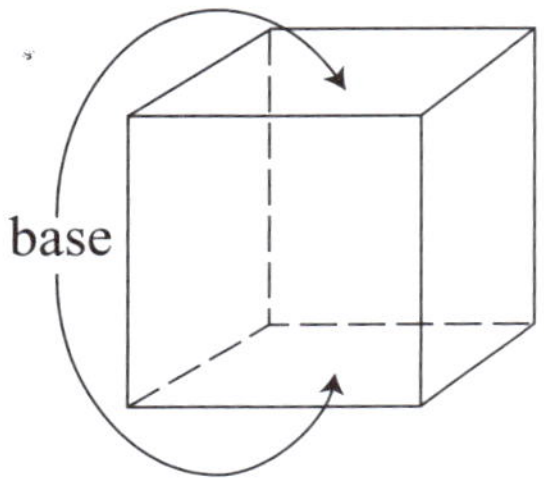 base	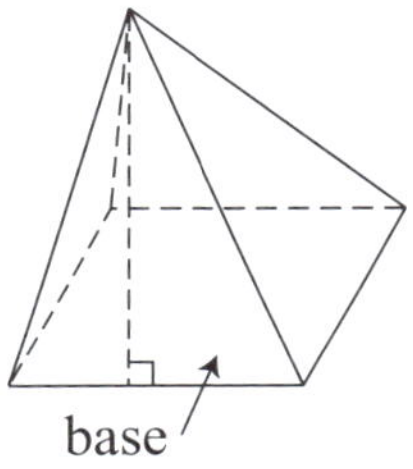 base	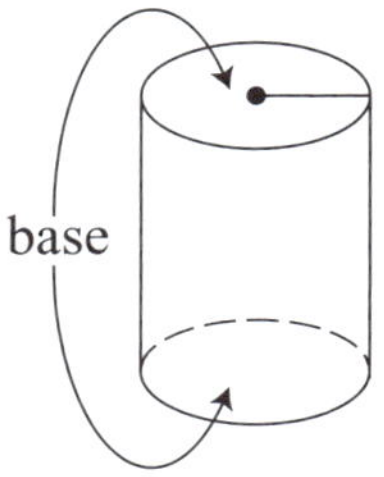 base	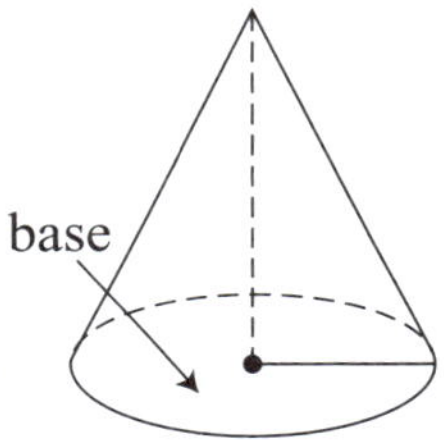 base	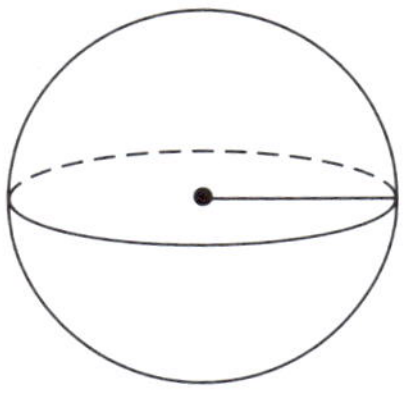
a pair of parallel polygon bases	one polygon base	a pair of parallel circular bases	one circular base	a set of points in space that are equidistant from a given point

Almost all three-dimensional objects can be represented by a two-dimensional figure called a **net**.

Prism	Pyramid	Cylinder	Cone	Sphere
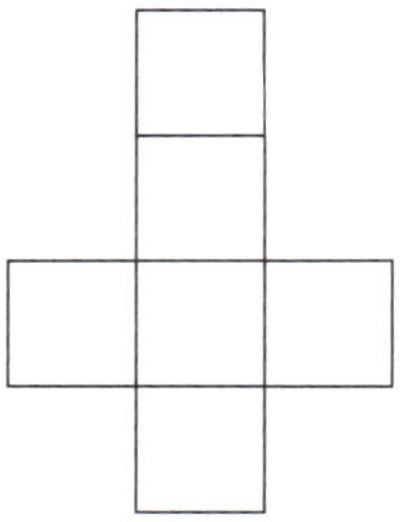	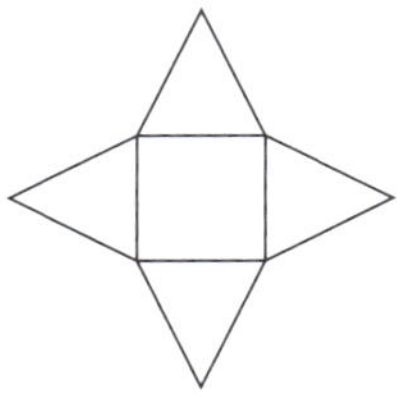	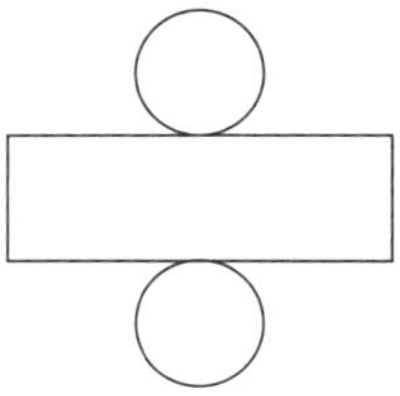	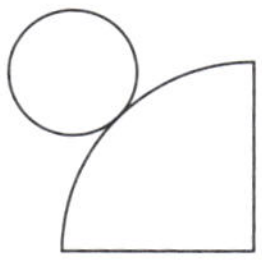	None

Solid figures are classified as polyhedrons or non-polyhedrons. A **polyhedron** is an enclosed solid figure with all flat surfaces. Prisms and pyramids are polyhedrons, but cylinders, cones, and spheres are not polyhedrons.

All polyhedrons are characterized by their faces, edges, and vertices. The flat surfaces, known as **faces**, are polygons. The **edges** are line segments formed by the intersection of two faces. The **vertices** are points formed by the intersection of three or more edges.

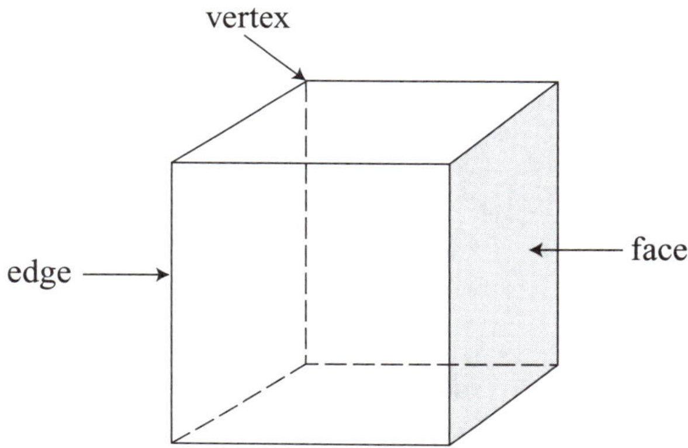

Regardless of the polyhedron, the number of faces (F), edges (E), and vertices (V) are related by Euler's Formula.

Euler's Formula

$$V + F - E = 2$$

Example:

A polyhedron has 4 faces and 4 vertices. Determine the number of edges.

Euler's Formula states that $V + F - E = 2$. Using substitution, $4 + 4 - E = 2$. Then, $8 - E = 2$ and $E = 6$. There are 6 edges in a polyhedron with 4 faces and 4 vertices. (It is a pyramid with a triangular base.)

Example:

Given the net below, determine the number of edges.

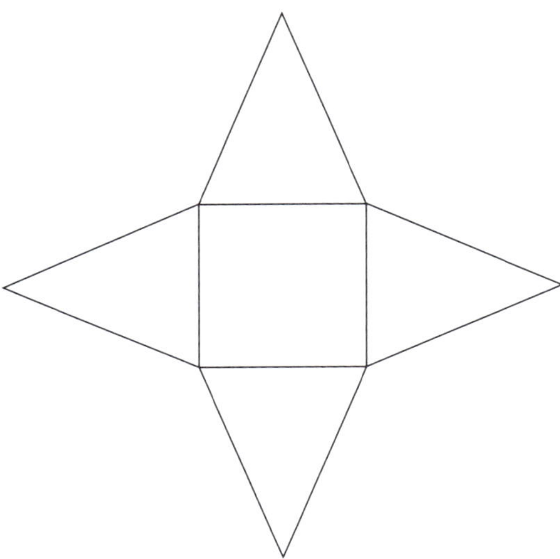

The edges are the line segments formed by the intersection of two faces. By counting the line segments in the given figure, there are 8 edges.

Note: Edges cannot be simply counted from a net because the intersection of some faces appear twce.

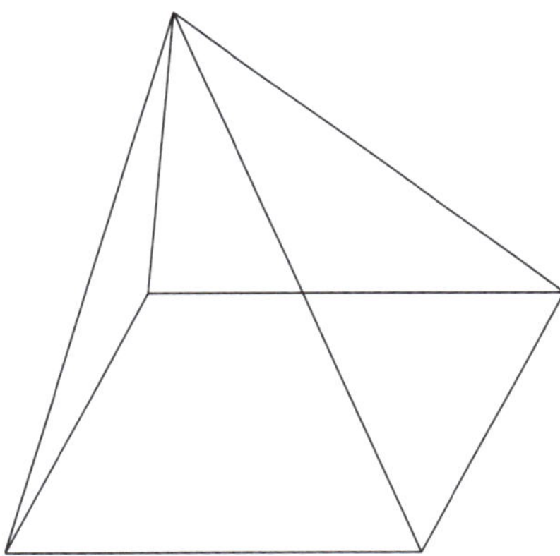

Example:

Determine the number of faces in an octahedron.

Regular polyhedrons are named for their number of congruent faces. An octahedron is a type of regular polyhedron. It is named with the prefix *octa-*, meaning 8. Therefore, an octahedron has 8 faces.

Example:

Which of the following is not a net of a cube?

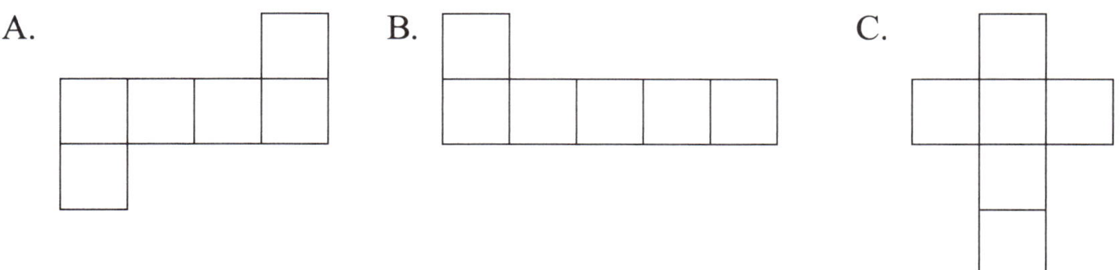

Figure B is not the net of a cube because if the figure were to folded, two faces would overlap and one of the sides would be open.

Note: They will all be equilateral triangles. It will also have 6 vertices and 12 edges.

Spheres

A **radius** of a sphere is a segment from the center to a point on the sphere. A **chord** is a segment with both endpoints on the sphere. A **diameter** is a type of chord that passes through the center of the sphere.

A **tangent** is a line that intersects a sphere in exactly one point.

A **great circle** is the largest circle that can be drawn on a sphere. It is a result of the intersection of a sphere and a plane that passes through the center of the sphere. Spheres have an infinite number of great circles and any diameter of any great circle coincides with the diameter of the sphere. All great circles divide a sphere into two congruent halves called hemispheres.

Example:

Match each of the following terms with the appropriate representation from the figure:

radius

chord

tangent

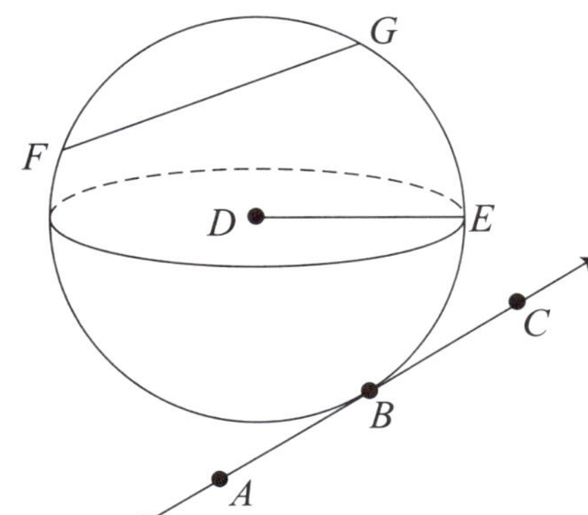

$\overline{DE}$ is a radius because the endpoints are the center of the sphere and a point on the sphere.

$\overline{FG}$ is a chord because both endpoints are on the sphere.

$\overleftrightarrow{AC}$ is a tangent because it intersects the sphere at a single point.

Example:

Determine the area of a great circle of the given sphere if $AB = 10$ in.

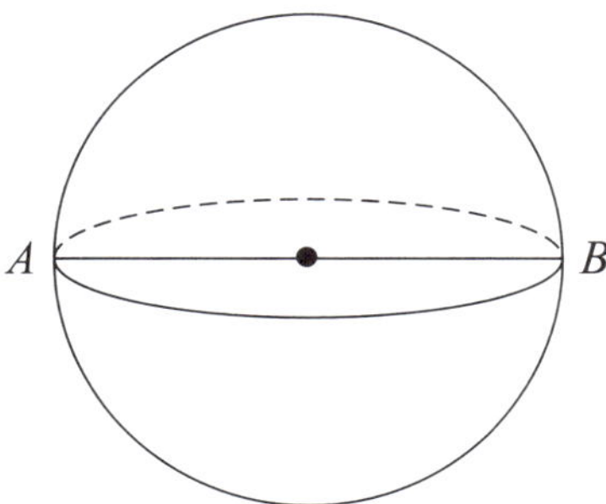

Since the diameter of a great circle coincides with the diameter of the sphere, that diameter of a great circle is 10 in. If the diameter is 10 in, the radius is 5 in. Using the area of a circle formula, $A = \pi r^2$, substitute the length of the radius.

$A = \pi r^2$

$A = \pi(5)^2$

$A = 25\pi$ square inches

Exercise 1

1. A polyhedron has 9 edges and 6 vertices. Determine the number of faces.

2. Given the net below, determine the number of edges.

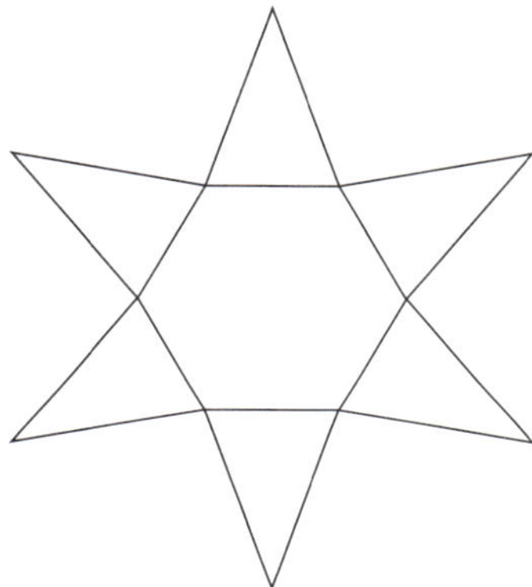

3. Given the polyhedron below, determine the number of vertices.

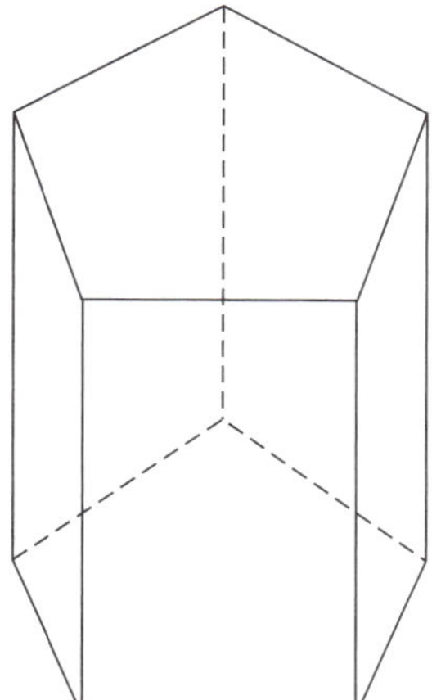

4. Identify two chords in the sphere given below.

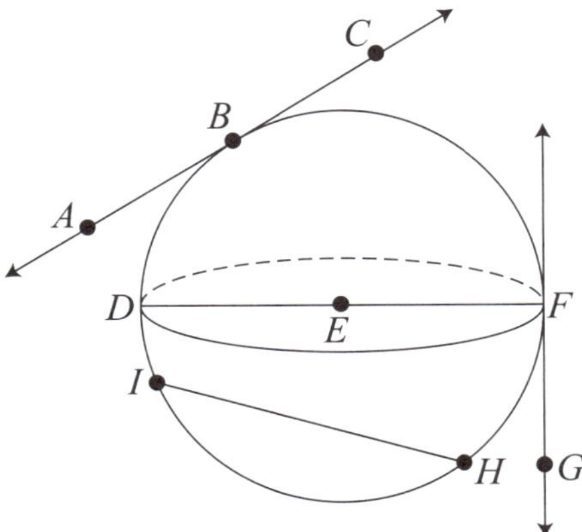

5. The circumference of a great circle of a sphere is approximately 50.24 feet. Determine the radius of the sphere.

Surface Area and Volume

All solid figures have a surface area and volume.

Surface area is the sum of the areas of all faces including the base(s). Surface area is measured in square units.

Volume is the amount of space inside a solid. Volume is measures in cubic units.

The table on the next page shows the formulas for common geometric solids.

		Surface Area	Volume
Prism	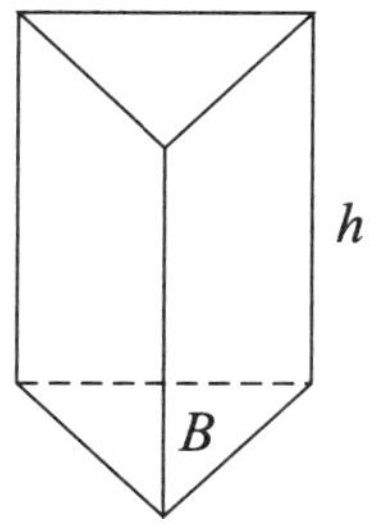	$SA = 2B + Ph$	$V = Bh$
Cylinder	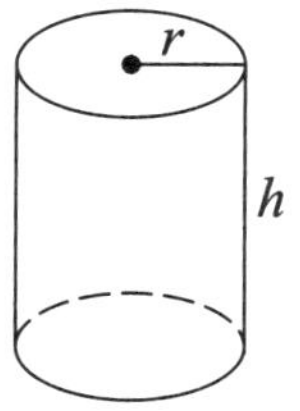	$SA = 2\pi r^2 + 2\pi rh$	$V = \pi r^2 h$
Pyramid	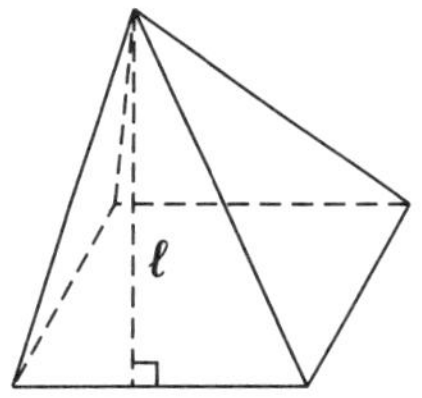	$SA = B + \frac{1}{2} Pl$	$V = \frac{1}{3} Bh$
Cone	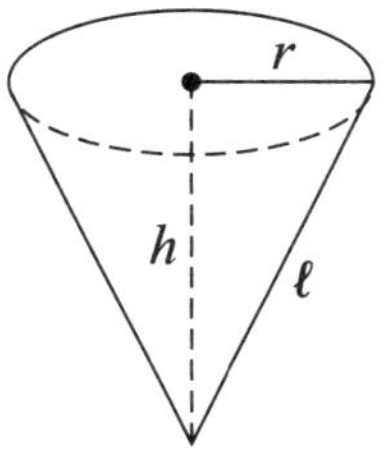	$SA = \pi r^2 + \pi rl$	$V = \frac{1}{3} \pi r^2 h$
Sphere	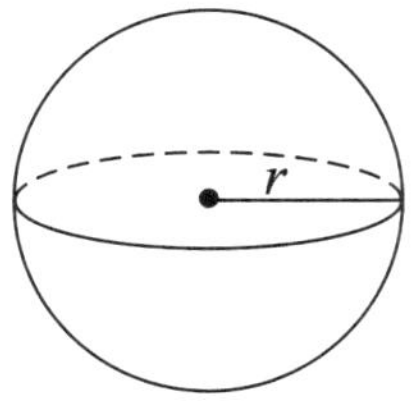	$SA = 4\pi r^2$	$V = \frac{4}{3} \pi r^3$

SA: surface area; V: volume; B: area of the base; P: perimeter of the base
l: slant height; h: height; r: radius

Example:

Determine the volume of the solid.

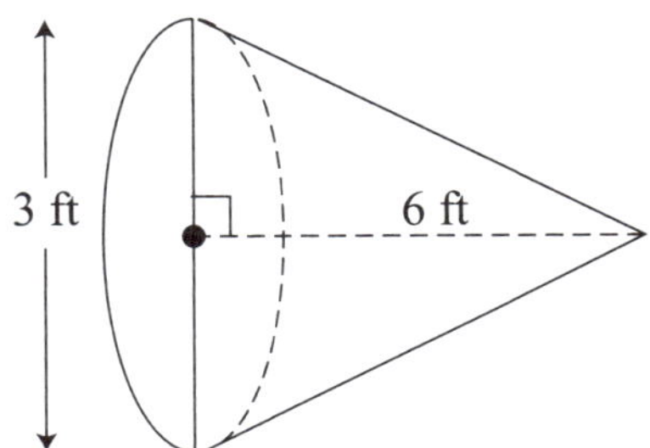

The solid given is a cone. The formula for the volume of a cone is $V = \frac{1}{3}\pi r^2 h$. The height of the cone is 6 ft and the radius $= \frac{1}{2}(3 \text{ ft}) = 1.5$ ft. Substituting into the formula:

$$V = \frac{1}{3}(\pi)(1.5 \text{ ft})^2(6 \text{ ft})$$

$$V = \frac{1}{3}(\pi)(2.25 \text{ ft}^2)(6 \text{ ft})$$

$$V = 4.5\pi \text{ ft}^3$$

Example:

Determine the surface area of a square pyramid with a height of 8 ft and a base with side length of 12 ft.

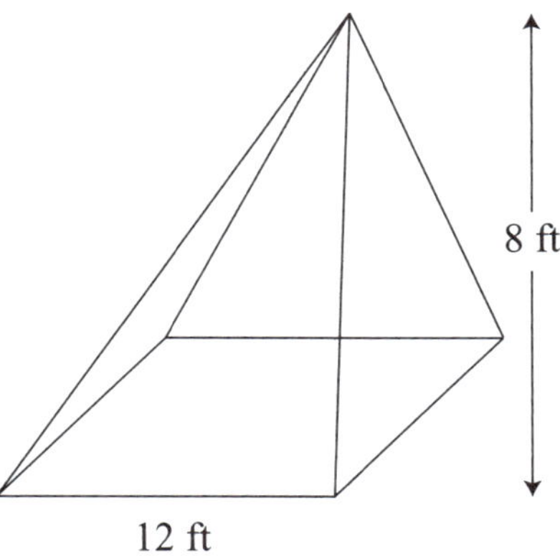

Note: Figure not drawn to scale

The formula for the surface area of a pyramid is $SA = B + \frac{1}{2}Pl$. Since the base is a square, the area of the base is $B = s^2 = (12 \text{ ft})^2 = 144 \text{ ft}^2$. The perimeter of the base is $P = 4s = 4(12 \text{ ft}) = 48$ ft.

The height of a square pyramid intersects the base at the center forming the right triangle below:

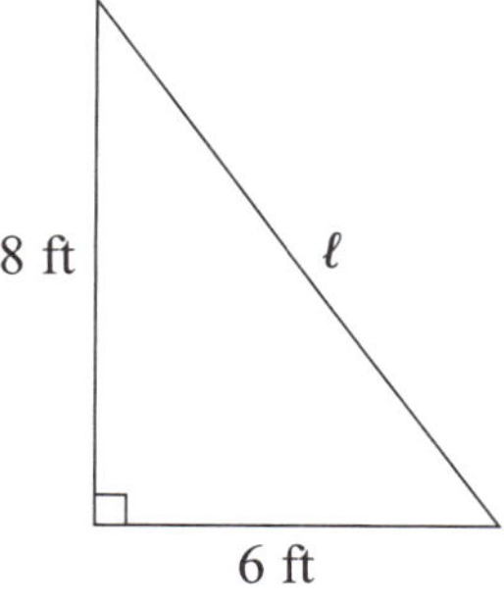

The slant height is found by using the Pythagorean Theorem.

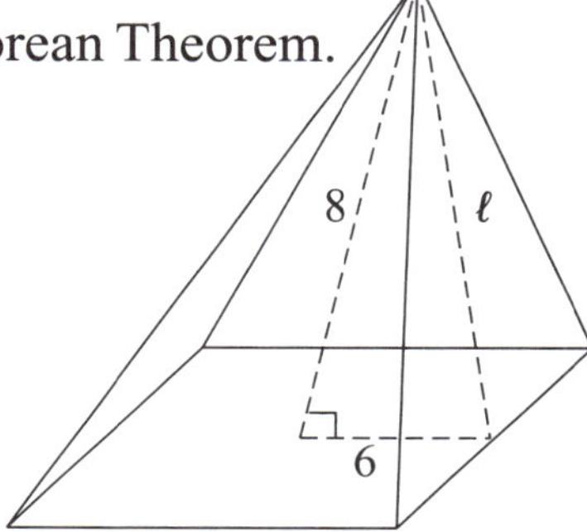

$a^2 + b^2 = c^2$

$6^2 + 8^2 = \ell^2$

$100 = \ell^2$

$10 = \ell$

The slant height is 10 ft. Substituting into the formula:

$$SA = B + \frac{1}{2}Pl$$

$$SA = (144 \text{ ft})^2 + \frac{1}{2}(48 \text{ ft})(10 \text{ ft})$$

$$SA = (144 \text{ ft}^2) + (240 \text{ ft}^2)$$

$$SA = 384 \text{ ft}^2$$

Example:

Determine the surface area of the solid.

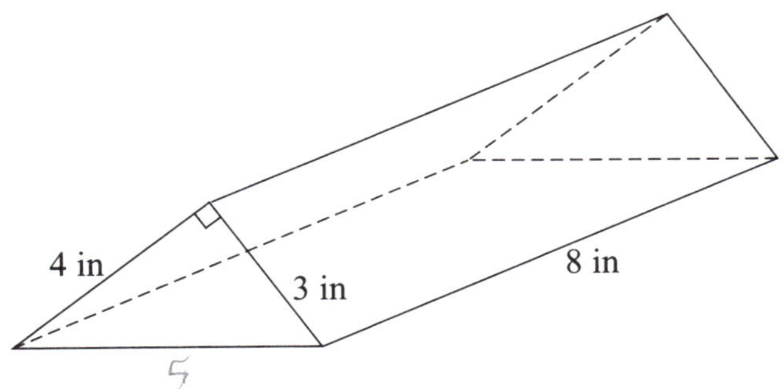

The solid is a prism with a triangular base; therefore, it is called a triangular prism. The formula for the surface area of a prism is $SA = 2B + Ph$. Since the base is a triangle, the area of the base is $B = \frac{1}{2}bh = \frac{1}{2}(3\text{ in})(4\text{ in}) = 6\text{ in}^2$. To find the perimeter of the base (P), the length of the third side needs to be determined. Since it is a right triangle, the Pythagorean Theorem can be used.

$$a^2 + b^2 = c^2$$

$$4^2 + 3^2 = c^2$$

$$25 = c^2$$

$$5 = c$$

The sides of the triangle measure 3 in., 4 in., and 5 in., so $P = 12$ in. Substituting into the formula:

$$SA = 2(6\text{ in})^2 + (12\text{ in})(8\text{ in})$$

$$SA = 12\text{ in}^2 + 96\text{ in}^2$$

$$SA = 108\text{ in}^2$$

Note: The height of the triangle is 4 and differs from the height of the prism which is 8. However, they both use the symbol h.

Exercise 2

1. Determine the volume of the solid.

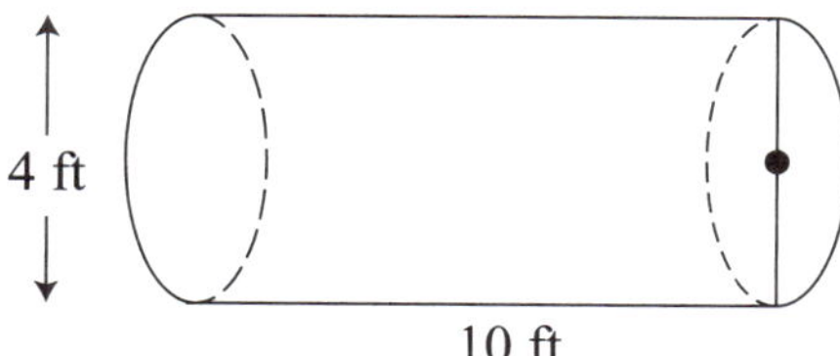

2. Determine the surface area of the solid.

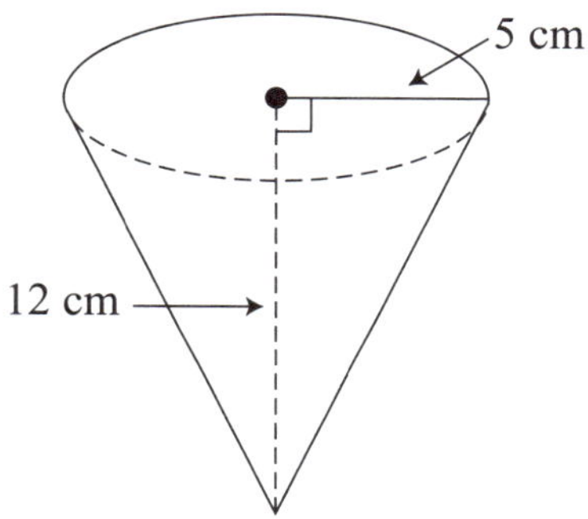

3. Determine the volume of a sphere with a diameter of 6 in.

4. The area of a great circle of a sphere is 25 π square meters. Determine the surface area of the sphere.

5. Determine a possible height and radius of a cylinder that results in a volume that measures between 110 yd^3 and 120 yd^3. *Hint:* Select an integer value less than 6 for *r*.

Congruent and Similar Solids

Two solids are considered **congruent solids** if they meet four criteria: (1) corresponding angles are congruent, (2) corresponding edges are congruent, (3) corresponding faces are congruent, and (4) volumes are equal.

Similar solids are solids that have exactly the same shape, but not necessarily the same size. Their corresponding angles are congruent. Their corresponding faces are similar. Two solids can be proven similar by comparing the ratios of the corresponding edge measures, known as the **scale factor**.

Example:

Find the scale factor for each pair of similar solids.

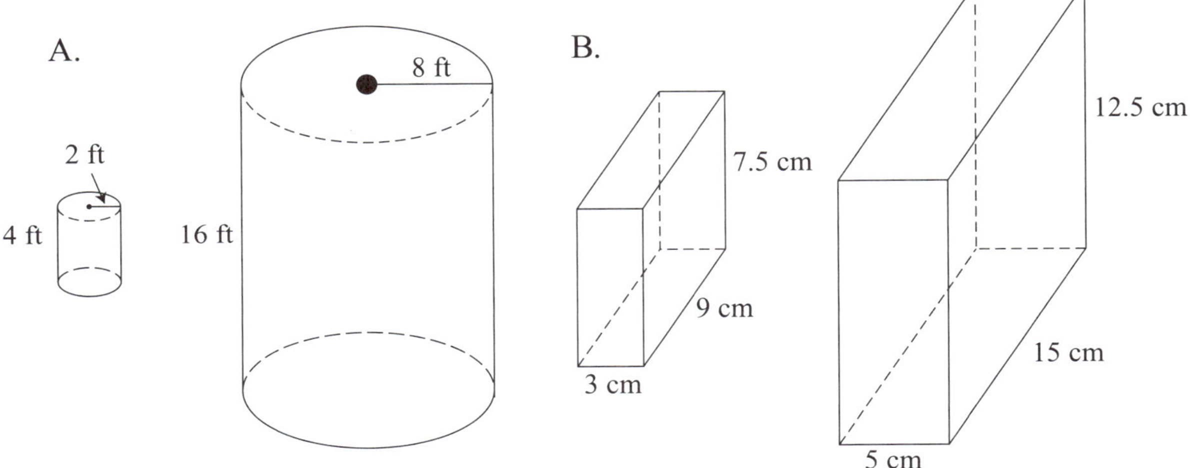

The similar solids in A have a scale factor of $\frac{1}{4}$. The heights are in ratio of $\frac{4}{16}$, which simplifies as $\frac{1}{4}$ and the radii are in ratio of $\frac{2}{8}$, which simplifies as $\frac{1}{4}$.

The similar solids in B have a scale factor of $\frac{3}{5}$. The lengths are in ratio of $\frac{3}{5}$, the widths are in ratio of $\frac{9}{15}$, which simplifies as $\frac{3}{5}$, and the heights are in ratio of $\frac{7.5}{12.5}$, which simplifies as $\frac{3}{5}$.

Example:

Determine if the solids are similar.

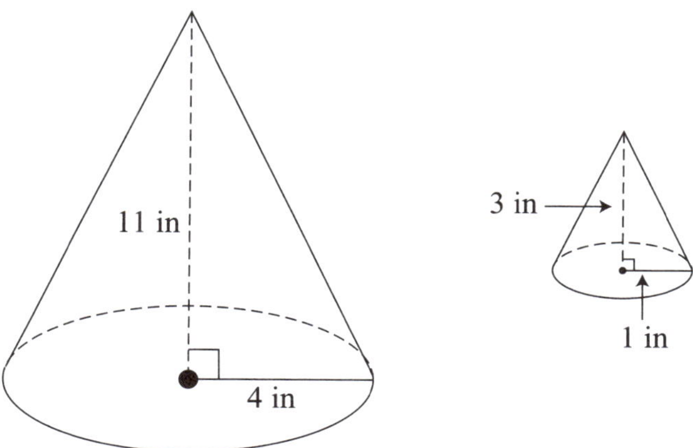

The ratio of the heights is $\frac{11}{3}$ and the ratio of the radii is $\frac{4}{1}$. Since $\frac{11}{3} \neq \frac{4}{1}$, the solids are not similar.

Effect of a Dilation on Surface Area and Volume

Prism A is a reduction of Prism B by a scale factor of $\frac{2}{3}$.

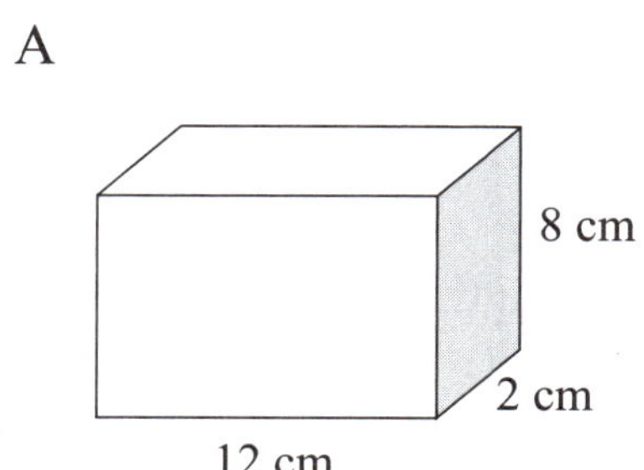

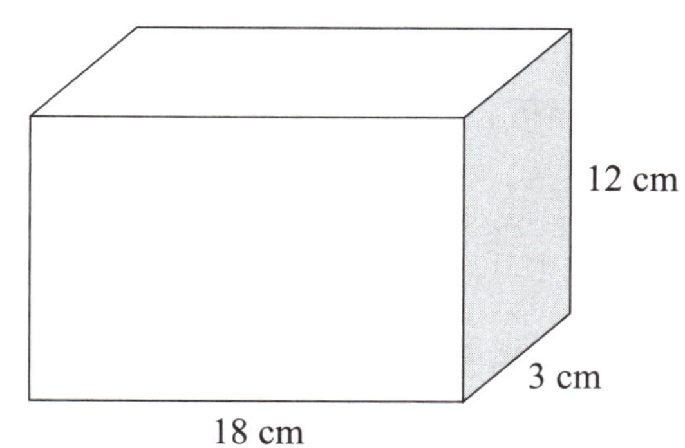

Surface area of Prism $A = 2B + Ph = 2(24) + (28)(8) = 272 \text{ cm}^2$

Surface area of Prism $B = 2B + Ph = 2(54) + (42)(12) = 612 \text{ cm}^2$

The ratio of the surface areas is $\frac{272}{612} = \frac{4}{9} = \frac{2^2}{3^2}$. In general, if two solids are similar with a scale factor of $\frac{a}{b}$, then the surface areas have a ratio of $\frac{a^2}{b^2}$.

Volume of Prism $A = Bh = (24)(8) = 192 \text{ cm}^3$

Volume of Prism $B = Bh = (54)\ (12) = 648 \text{ cm}^3$

The ratio of the volumes is $\frac{192}{648} = \frac{8}{27} = \frac{2^3}{3^3}$. In general, if two solids are similar with a scale factor of $\frac{a}{b}$, then the volumes have a ratio of $\frac{a^3}{b^3}$.

Example:

Determine the ratio of the volumes of two similar solids that have a surface area ratio of 9:16.

Surface areas have a ratio of $\frac{a^2}{b^2}$. Therefore, $a^2 = 9$ and $b^2 = 16$ then $a = 3$ and $b = 4$. Volumes have a ratio of $\frac{a^3}{b^3}$, so the ratio of the volumes is $\frac{3^3}{4^3}$ which is $\frac{27}{64}$.

Example:

A pyramid with a surface area of 20 in^2 is enlarged by a scale factor of 2. What is the surface area of the new pyramid?

The scale factor of the pyramids is $\frac{2}{1}$, meaning their lengths have a $\frac{2}{1}$ ratio, therefore the surface areas are in a ratio of $\frac{2^2}{1^2} = \frac{4}{1}$. The surface area of the original pyramid is 20 in^2, so the following proportion can be used to solve for the surface area of the enlarged pyramid:

$$\frac{4}{1} = \frac{x}{20}$$

$$x = 80$$

The surface area of the enlarged pyramid is 80 in^2.

Exercise 3

1. Determine if the two figures are similar, congruent or neither.

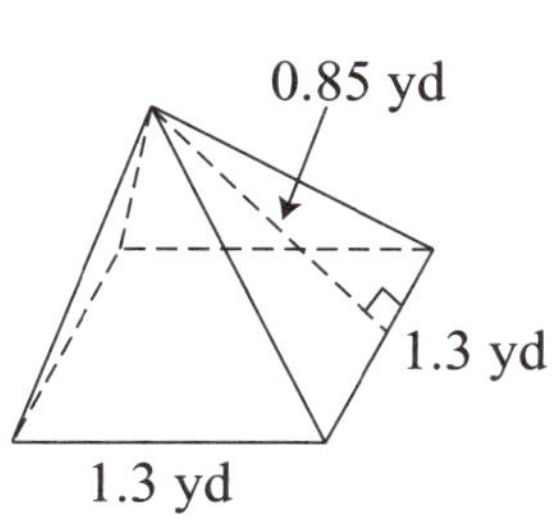

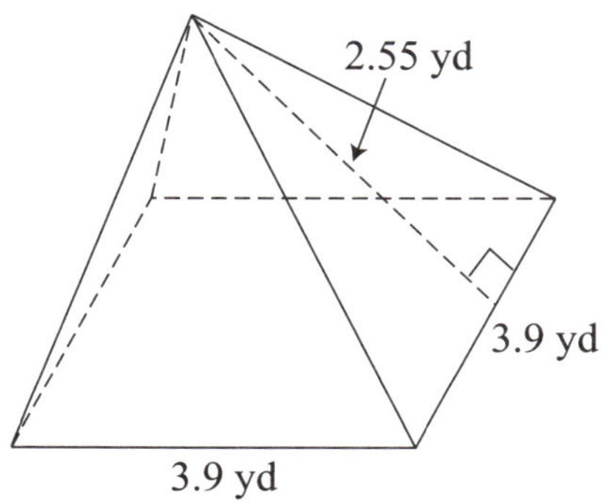

2. Find the scale factor for the pair of similar solids.

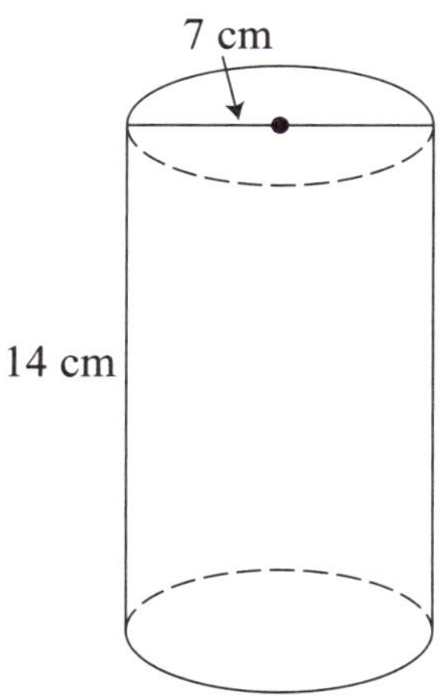

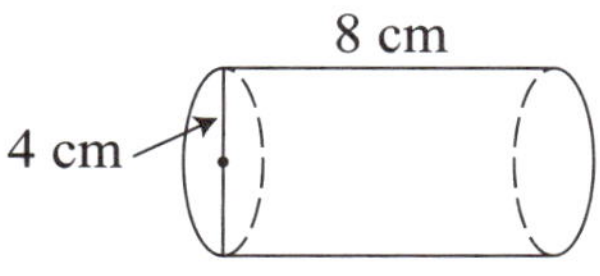

3. A prism with a surface area of 36 ft^2 is reduced to a prism with a surface area of 16 ft^2. What is the ratio of the volume of the original prism to the reduced prism?

End-of-Chapter Quiz

1. **Given that a polyhedron has 7 faces and 15 edges, which of the following is the number of vertices?**

   A. 7

   B. 10

   C. 15

   D. 20

2. **A pentagonal pyramid has _____ edges.**

   A. 5

   B. 6

   C. 10

   D. 15

3. **Draw the net of a cylinder.**

4. Which of the following best describes $\overleftrightarrow{AB}$?

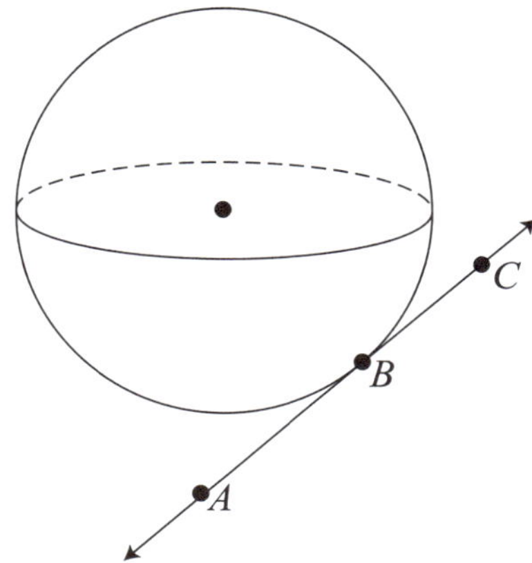

A. tangent

B. chord

C. radius

D. diameter

5. The circumference of a sphere's great circle is 20π meters. What is the surface area of the sphere?

A. 100π square meters

B. 400π square meters

C. 1000π square meters

D. 1600π square meters

6. **Determine the surface area of the cone.**

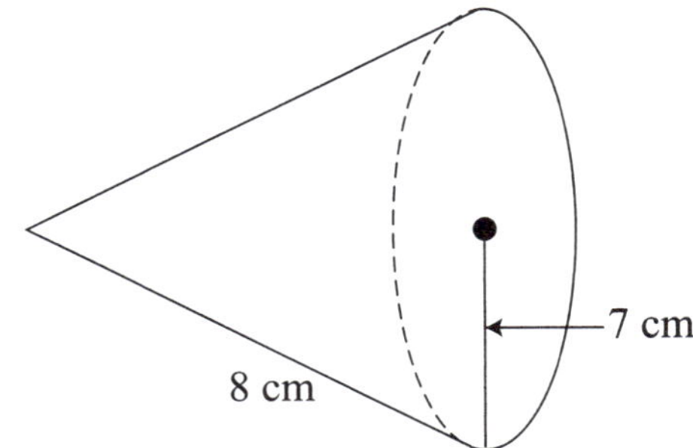

A. 62π square centimeters

B. 70π square centimeters

C. 97π square centimeters

D. 105π square centimeters

7. **Determine the volume of the pyramid.**

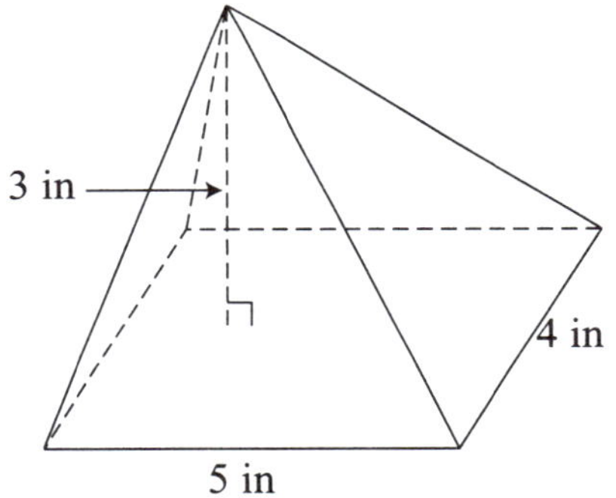

A. 16 cubic inches

B. 20 cubic inches

C. 25 cubic inches

D. 60 cubic inches

8. **Determine the surface area of the regular pentagonal prism.**

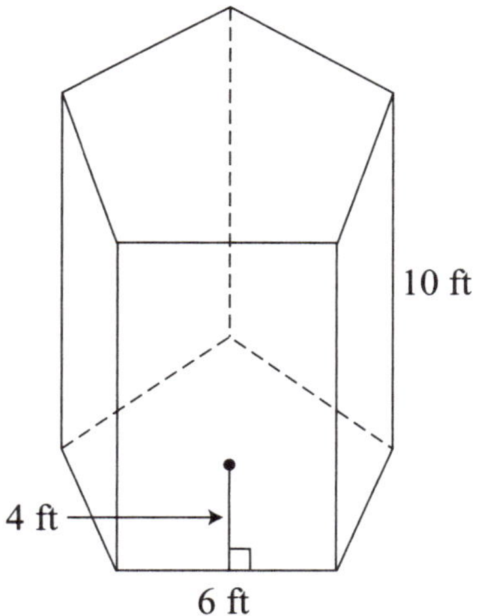

A. 309 square feet

B. 369 square feet

C. 420 square feet

D. 450 square feet

9. **If water fills 75% of the tank below, what is the approximate volume of the water?**

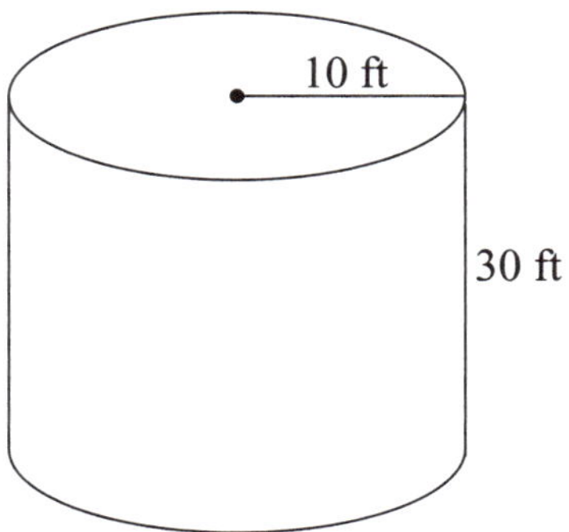

A. 2355 ft^3

B. 7068 ft^3

C. 9425 ft^3

D. 12560 ft^3

10. How many rectangular prisms with dimensions 2 in × 3 in × 2 in can fit in a box that is 1 ft × 2 ft × 1 ft?

A. 144

B. 216

C. 288

D. 432

11. Jenni is going to use the wooden box below to store her DVDs. Since her favorite color is yellow, she is going to paint the exterior of the box with yellow paint. If the wooden box has no top, how many square inches (in^2) does Jenni need to paint?

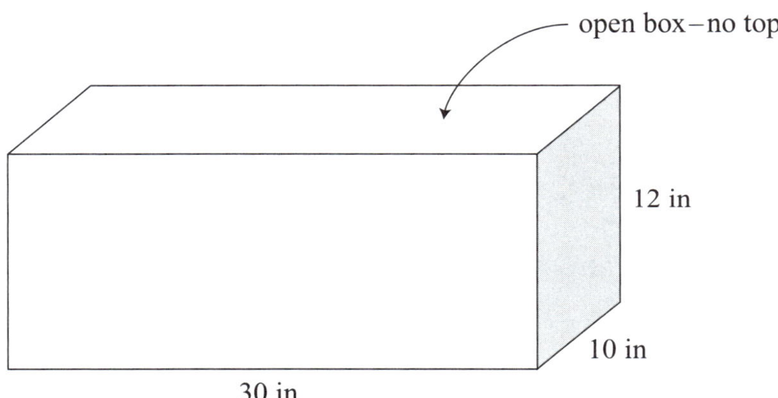

A. 52

B. 1260

C. 1560

D. 3600

12. Which of the following statements describes the relationship between the volume of cylinder A and the volume of cylinder B?

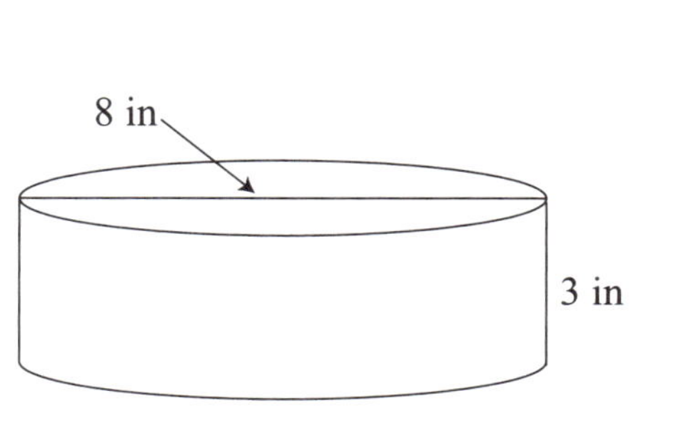

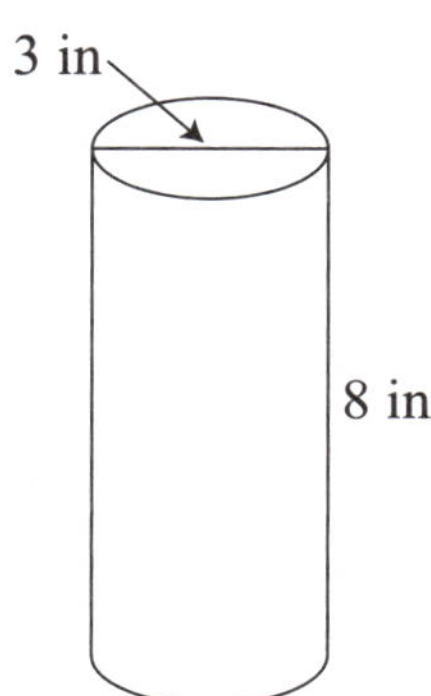

A. The volume of cylinder A is less than the volume of cylinder B.

B. The volume of cylinder A is greater than the volume of cylinder B.

C. The volume of cylinder A is equal to the volume of cylinder B.

D. The volume of cylinder A is double the volume of cylinder B.

13. If all the dimensions of a cone are tripled, which of the following best describes the relationship of the surface area of the new cone and the surface area of the original cone?

A. The surface area of the new cone is 3 times the surface area of the original cone.

B. The surface area of the new cone is 6 times the surface area of the original cone.

C. The surface area of the new cone is 9 times the surface area of the original cone.

D. The surface area of the new cone is 27 times the surface area of the original cone.

14. The two cylinders below are similar. The volume of cylinder A is 27 cm^3 and the volume of cylinder B is 1 cm^3.

Cylinder A

Cylinder B

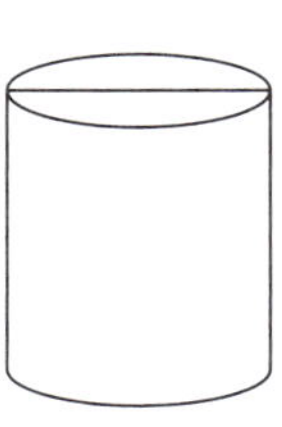

What is the radius of cylinder B?

A. 3 cm

B. 6 cm

C. 9 cm

D. 12 cm

15. The rectangular prisms below are similar.

Prism B

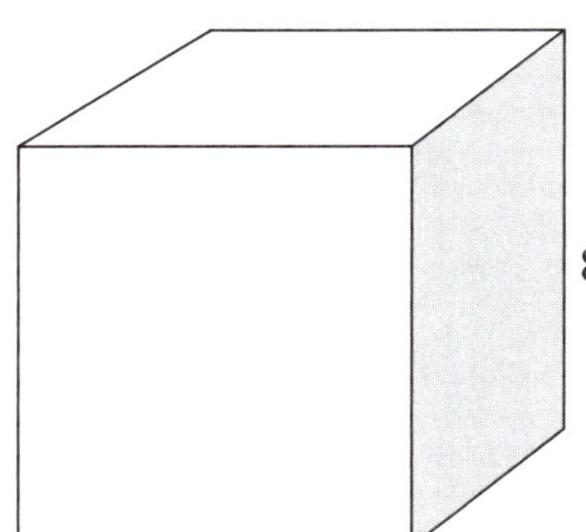

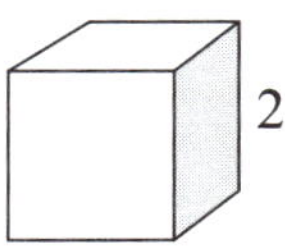

Which of the following is the ratio of the surface area of Prism A to the surface area of Prism B?

A. $\frac{1}{4}$

B. $\frac{16}{1}$

C. $\frac{1}{16}$

D. $\frac{4}{1}$

Answers to Exercise 1

1. 5

Using Euler's Formula, $V + F - E = 2$, substitute the number of edges and vertices.

$6 + F - 9 = 2$

$F - 3 = 2$

$F = 5$

The polyhedron has 5 faces.

2. 18

The edges are line segments formed by the intersection of 2 faces. The polyhedron formed by the net has 18 edges.

3. 10

The vertices are points formed by the intersection of 3 or more edges. The polyhedron has 10 vertices.

4. $\overline{DF}$ and $\overline{IH}$

Chords are line segments whose endpoints are on the sphere. $\overline{DF}$ and $\overline{IH}$ are chords.

5. 8

$C = 2\pi r$

$50.24 = 2(3.14)r$

$50.24 = 6.28r$

$8 = r$

Answers to Exercise 2

1. 40 cubic feet

If diameter is 4 ft, radius is 2 ft.

$V = \pi r^2 h$

$V = \pi(2)^2(10)$

$V = \pi(4)(10)$

$V = 40\pi$ cubic feet

2. 90π square centimeters

First, to determine the slant height, use the Pythagorean Theorem.

$$a^2 + b^2 = c^2$$

$$5^2 + 12^2 = l^2$$

$$25 + 144 = l^2$$

$$169 = l^2$$

$$13 = l$$

Substitute into the surface area formula,

$$SA = \pi r^2 + \pi r l$$

$$SA = \pi(5)^2 + \pi(5)(13)$$

$$SA = 25\pi + 65\pi$$

$$SA = 90\pi \text{ square centimeters}$$

3. 36π cubic inches

If diameter is 6 in, radius is 3 in.

$$V = \frac{4}{3}\pi r^3$$

$$V = \frac{4}{3}\pi(3)^3$$

$$V = \frac{4}{3}\pi(27)$$

$$V = 36\pi \text{ cubic inches}$$

4. 100π square meters

First, calculate the radius from the given area.

$A = \pi r^2$

$25\pi = \pi r^2$

$25 = r^2$

$5 = r$

The radius of the sphere equals the radius of the great circle, and is 5 meters. Substitute into the formula for surface area of a sphere.

$SA = 4\pi r^2$

$SA = 4\pi(5)^2$

$SA = 4\pi(25)$

$SA = 100\pi$ square meters

5. 4 yards

To find a radius and height that result in a volume between 110 yd^3 and 120 yd^3, write the following inequality: $110 < \pi r^2 h < 120$

Let $r = 3$.

$110 < \pi r^2 h < 120$

$110 < \pi(3)^2 h < 120$

$110 < \pi(9)h < 120$

$110 < 28.26h < 120$

$3.89 < h < 4.25$

If the radius is 3 yards, a possible height is 4 yards.

Answers to Exercise 3

1. The two figures are similar

The ratio of the slant heights is $\frac{0.85}{2.55} = \frac{1}{3}$

The ratio of the sides of the base is $\frac{1.3}{3.9} = \frac{1}{3}$

Since the ratios are equal, the figures are similar.

2. $\frac{7}{4}$

The ratio of the heights is $\frac{14}{8} = \frac{7}{4}$

The ratio of the diameters is $\frac{7}{4}$

Therefore, the scale factor is $\frac{7}{4}$

3. $\frac{27}{8}$

The ratio of surface areas of similar solids is $\frac{a^2}{b^2}$. So, $\frac{a^2}{b^2} = \frac{36}{16}$ and $\frac{a}{b} = \frac{6}{4} = \frac{3}{2}$.

The ratio of volumes for similar solids is $\frac{a^3}{b^3}$ so $\frac{a^3}{b^3} = \frac{3^3}{2^3} = \frac{27}{8}$.

Answers to the Chapter Quiz

1. B

$V + F - E = 2$

$V + 7 - 15 = 2$

$V - 8 = 2$

$V = 10$

2. C

A pentagon has 5 sides, so the base of the pyramid has 5 edges. Another 5 edges connect each vertex of the base to the vertex of the pyramid.

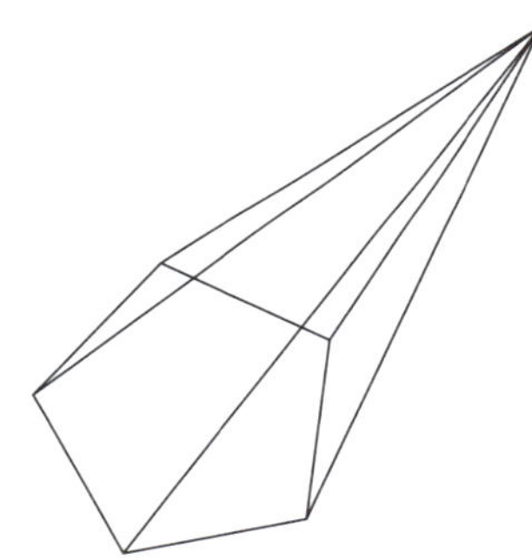

3.

4. A

$\overleftrightarrow{AB}$ is a line that intersects the sphere at one point; therefore, it is a tangent.

5. B

$C = 2\pi r$

$20\pi = 2\pi r$

$10 = r$

$SA = 4\pi r^2$

$SA = 4\pi(10)^2$

$SA = 4\pi(100)$

$SA = 400\pi$ square meters

6. D

$SA = \pi r^2 + \pi rl$

$SA = \pi(7)^2 + \pi(7)(8)$

$SA = 49\pi + 56\pi$

$SA = 105\pi$ square centimeters

7. B

$$V = \frac{1}{3}Bh$$

$$V = \frac{1}{3}(20)(3)$$

$V = 20$ cubic inches

8. C

The base of the prism is a regular pentagon consisting of 5 congruent triangles with base = 6 ft and height = 4 ft. The area of one triangle = $\frac{1}{2}bh = \frac{1}{2}(6)(4) = 12$ square feet. The area of the pentagon base is 5 times the area of the triangle, B = 5(12) = 60 square feet.

$SA = 2B + Ph$

$SA = 2(60) + 30(10)$

$SA = 120 + 300$

$SA = 420$ square feet

9. B

$V = \pi r^2 h$

$V = \pi(10)^2(30)$

$V = 3000\pi$

$V \approx 9425$ cubic feet

The volume of the entire tank is 9425 cubic feet. The water is 75% of the volume of the tank. Volume of water $\approx (0.75)(9425) \approx 7068$ cubic feet.

10. B

Let the three dimensions of each rectangular prism be represented by length, width, and height, respectively. Divide 1 ft (12 in.) by 2 in. to get 6. Divide 2 ft (24 in.) by 3 to get 8. In order to fit the smaller rectangular prisms into the larger rectangular prism (box), we can only use the *minimum* value of the pair of 6 and 8, which is 6. Then $6^3 = 216$ represents the *maximum* number of rectangular prisms that can fit into the box. (Note that there will be "unoccupied" space within the box.)

11. B

To determine the area that will be painted, find the area of each side except the top and add them together.

Area of bottom = 30(10) = 300

Area of front = 30 (12) = 360

Area of back = 30(12) = 360

Area of left side = 10(12) = 120

Area of right side = 10(12) = 120

Area to be painted = 300 + 2(360) + 2(120) = 1260 square inches

12. B

Cylinder A

$V = \pi r^2 h$

$V = \pi(4)^2(3)$

$V = 48\pi$ cubic inches

Cylinder B

$V = \pi r^2 h$

$V = \pi(1.5)^2(8)$

$V = 18\pi$ cubic inches

13. C

If the dimensions of a cone are tripled, the scale factor is $\frac{3}{1}$. The ratio of the surface areas is equal to the scale factor squared: $\left(\frac{3}{1}\right)^2 = \frac{9}{1}$. The surface area of the new cone is 9 times the surface area of the original cone.

14. A

The ratios of the volumes of similar solids is $\frac{a^3}{b^3}$. If $\frac{a^3}{b^3} = \frac{27}{1}$ then $\frac{a}{b} = \frac{3}{1}$. The ratio of the diameters is $\frac{3}{1}$. Therefore, $\frac{3}{1} = \frac{18}{x}$, so $x = 6$. The diameter of cylinder B is 6 cm, so the radius is 3 cm.

15. B

The ratio of the heights is $\frac{8}{2} = \frac{4}{1}$. The ratio of the surface areas is $\frac{4^2}{1^2} = \frac{16}{1}$.

Chapter 7

Mathematical Reasoning and Problem Solving

Your Goals for Chapter 7

1. You should be able distinguish between undefined terms, definitions, postulates, and theorems.
2. You should be able to apply a variety of problem solving strategies and select an appropriate strategy based on the context of the question.
3. You should be able to determine if a solution is reasonable.
4. You should be able to make and justify conjectures.
5. You should be able to write geometric proofs in a variety of ways.

Standards

The following standards are assessed on Florida's Geometry End-of-Course exam either directly or indirectly:

09MA.912.G.8.1: (High) Analyze the structure of Euclidean geometry as an axiomatic system. Distinguish between undefined terms, definitions, postulates, and theorems.

MA.912.G.8.2: (Moderate) Use a variety of problem-solving strategies, such as drawing a diagram, making a chart, guess-and-check, solving a simpler problem, writing an equation, and working backwards.

MA.912.G.8.3: (Moderate) Determine whether a solution is reasonable in the context of the original situation.

MA.912.G.8.4: (High) Make conjectures with justifications about geometric ideas. Distinguish between information that supports a conjecture and the proof of a conjecture.

MA.912.G.8.5: (High) Write geometric proofs, including proofs by contradiction and proofs involving coordinate geometry. Use and compare a variety of ways to present deductive proofs, such as flow charts, paragraphs, two-column, and indirect proofs.

The Axiomatic Structure of Euclidean Geometry

An axiomatic system consists of undefined terms, definitions, postulates (axioms), and theorems.

Undefined terms are words that are not formally defined. Euclidean Geometry contains three undefined terms: points, lines, and planes. The objects can be visualized, but cannot be defined.

A **definition** is a description of a new word using previously known words or terms.

A **postulate** is a true statement that does not require a proof. Postulates are used to derive logical statements that can be used to solve a problem.

A **theorem** is a statement that must be proven. The proof consists of previously proven theorems or postulates.

Example:

Classify each of the following as an undefined term, defined term, postulate, or theorem.

A. angle

B. side-side-side congruence

C. point

D. Pythagorean Theorem

A. An angle is a defined term because it is defined with the use of known words or terms. The definition of an angle is a figure formed by two rays that have a common endpoint.

B. The side-side-side congruence is a postulate because it does not require a proof. Instead, it is used to prove whether two given triangles are congruent.

C. As stated, a point is one of the undefined terms of Euclidean Geometry. It cannot be defined.

D. The Pythagorean Theorem is a theorem. The theorem was not assumed, it was proven through the use of other theorems and postulates.

Example:

Determine if the statement is a postulate or a theorem.

A. If two planes intersect, then their intersection is a line.

B. Vertical angles are congruent.

The statement given in A is a postulate. Assume that the walls in your room are planes. In the corner where they intersect, they form a line. That is the only intersection of those two planes. Since the statement does not require a proof, it is a postulate.

The statement given in B is a theorem. The following is a proof of the theorem:

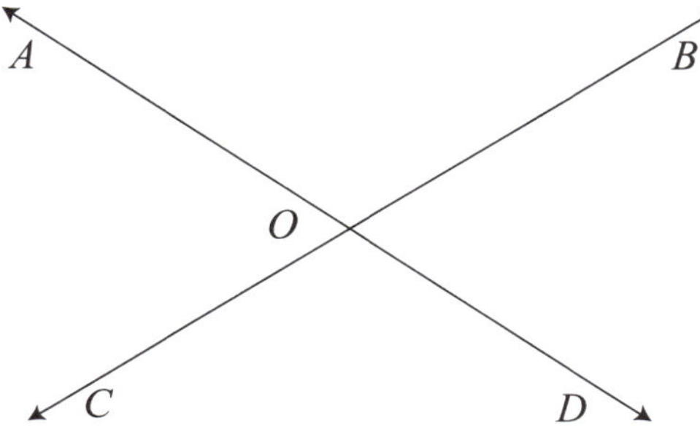

$\angle BOD$ is supplementary to $\angle AOB$ because they form a straight angle

$\angle AOC$ is supplementary to $\angle AOB$ because they form a straight angle

$\angle BOD$ is congruent to $\angle AOC$ because two angles that are supplementary to the same third angle are congruent.

Since the statement "vertical angles are congruent" was proven through the use of other theorems and postulates, it is a theorem.

Problem Solving Strategies

To be an effective problem solver, it is important to have numerous strategies and be able to identify the best strategy for a given situation. Some of the most popular problem solving strategies are:

- Draw a diagram
- Make a chart
- Guess and check
- Solve a simpler problem
- Write an equation

Example:

Identify a strategy to solve the problems. Apply the strategy to obtain the solution.

A. The area of a rectangle is 36 square inches. If the dimensions are whole numbers, what are the dimensions that result in the shortest perimeter?

Since there are a limited number of rectangles that fit the criteria, it is a good strategy to make a chart.

length	width	area	perimeter
1 in.	36 in.	36 in^2	74 in.
2 in.	18 in.	36 in^2	40 in.
3 in.	12 in.	36 in^2	30 in.
4 in.	9 in.	36 in^2	26 in.
6 in.	6 in.	36 in^2	24 in.

Based on the information in the chart, a 6 inches (in.) × 6 in. rectangle would have the shortest perimeter.

B. Jackson has a picture that is 5 in. × 7 in. He put the picture in a frame that is 3 inches wide on all sides. What is the area of the entire picture frame?

A diagram can be used to represent this problem and will make it easier to solve the problem.

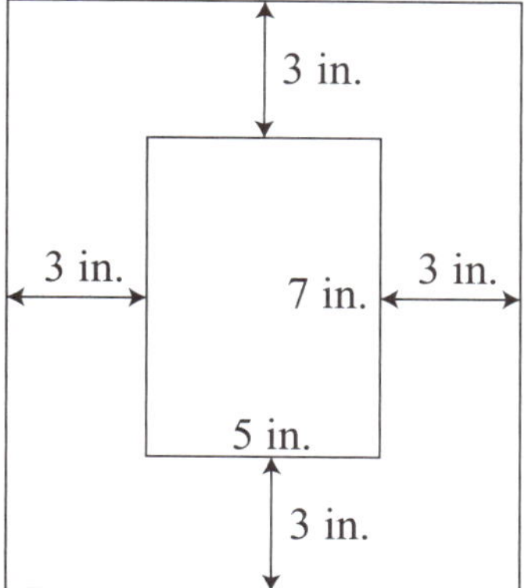

Based on the diagram, the length of the frame is 3 in. + 5 in. + 3 in. = 11 in. and the width of the frame is 3 in. + 7 in. + 3 in. = 13 in. The area of the frame is (11 in.)(13 in.) = 143 in^2.

Reasoning and Proof

Inductive reasoning uses a set of examples to arrive at a conclusion. The concluding statement based on inductive reasoning is called a **conjecture**. The conjecture is believed to be true based on the examples, but has yet to be proven.

Example:

The teacher posts the following information on the board:

The sun came up this morning.

The sun came up yesterday.

The sun came up the day before that.

Based on these examples, make a conjecture.

By using inductive reasoning, it can be concluded that the sun comes up every day. The examples given support the conjecture that was made.

Example:

Make a conjecture about the relationship. Provide examples that support your conjecture.

- the relationship between the areas of a square with side s and a rectangle with sides $2s$ and $3s$

Apply inductive reasoning by giving a set of examples

1) Assume $s = 1$, then $2s = 2$ and $3s = 3$.

Area of the square = 1 square unit

Area of the rectangle = 6 square units

2) Assume $s = 2$, then $2s = 4$ and $3s = 6$.

Area of the square = 4 square units

Area of the rectangle = 24 square units

3) Assume $s = 3$, then $2s = 6$ and $3s = 9$.

Area of the square = 9 square units

Area of the rectangle = 54 square units

Looking at the examples, the area of the rectangle appears to be 6 times as large as the area of the square.

Conjecture: the area of a rectangle with sides $2s$ and $3s$ is 6 times as large as the area of a square with side s.

Conjectures are made based on a set of examples. To show that a conjecture is true for all possible examples, a set of logical statements called a proof must be provided. However, to show that a conjecture is not true, only one false example must be found. This false example is called a **counterexample**.

Example:

Provide a counterexample to show that the conjecture is false.

All triangles are similar figures.

Counterexample:

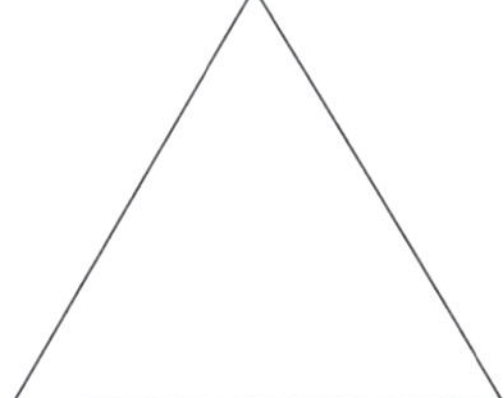

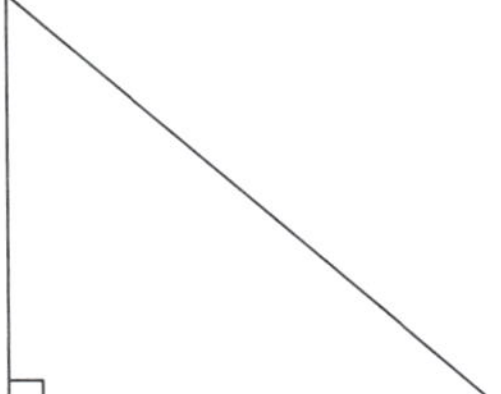

In the example given, the second triangle has a right angle and the first triangle does not. Therefore, the triangles are not similar. Similar figures must have corresponding congruent angles.

Deductive reasoning makes conclusions based on things assumed to be true. If the facts used to support a characteristic are true, then the characteristic itself is proven to be true.

In the most basic sense, the following is considered deductive reasoning:

Mr. Quack is a duck.

All ducks can swim.

Therefore, Mr. Quack can swim.

Since facts are provided that Mr. Quack is a duck and that all ducks can swim, it can be concluded that Mr. Quack can swim.

There are various ways to present deductive proofs, such as flow-charts, paragraphs, two-column, and indirect proof. Flow-charts, paragraphs, and two-column proofs are different formats to organize the step-by-step process of a proof. An **indirect proof** differs in that it is a type of proof in which a statement is assumed false. The assumption leads to an impossibility, thus proving the statement true. (Examples of indirect proof are provided in Chapter 8.)

Example:

Given: $\angle 1$ and $\angle 2$ are supplementary.

$\angle 2$ and $\angle 3$ are supplementary.

Prove: $\angle 1 \cong \angle 3$

Write a paragraph proof.

If $\angle 1$ and $\angle 2$ are supplementary, then the sum of their measures is 180°. If $\angle 2$ and $\angle 3$ are supplementary, then the sum of their measures is 180°. This means that the sum of $\angle 1$ and $\angle 2$ and the sum of $\angle 2$ and $\angle 3$ are equal by substitution. Using the reflexive property, $m\angle 2 = m\angle 2$, so $m\angle 1 = m\angle 3$ by the subtraction property. Definition of congruent angles proves that $\angle 1 \cong \angle 3$.

Example:

Given that figure *WXYZ* is a quadrilateral, Use a two-column proof to prove that it is a parallelogram.

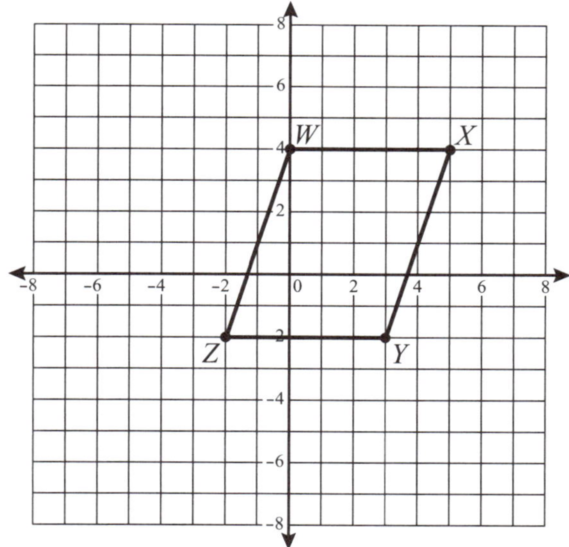

Statements	Reasons
1. slope of $WX = \dfrac{4-4}{0-5} = 0$	1. definition of slope
2. slope of $YZ = \dfrac{-2-(-2)}{-2-3} = 0$	2. definition of slope
3. slope of $WZ = \dfrac{4-(-2)}{0-(-2)} = 3$	3. definition of slope
4. slope of $XY = \dfrac{4-(-2)}{5-3} = 3$	4. definition of slope
5. $\overline{WX} \parallel \overline{YZ}$, $\overline{WZ} \parallel \overline{XY}$	5. parallel lines have equal slopes
6. *WXYZ* is a parallelogram	6. definition of a parallelogram

End-of-Chapter Quiz

1. Which of the following is a defined term?

- **A.** Parallelogram
- **B.** Plane
- **C.** Supplementary angles
- **D.** Both A and C

2. Determine if the statement is a postulate or theorem. Explain your reasoning.

Two lines intersect at one point

3. The circumference of a circle is 8π. Max states that the radius of the circle is 8. Is his answer reasonable? Explain.

4. $\angle ABC$ and $\angle DEF$ are complementary angles. If $m\angle ABC = 30°$, then $\angle DEF$ must be an obtuse angle. Is this reasonable? Explain.

5. The carnival ferris wheel is a circle with a diameter of 24 feet. It has 12 equally spaced seats that rotate clockwise. If Jonah is in the first seat, how far will he travel to get to the location of seat 5?

6. Juliann stated that all rectangles are squares. Provide a counterexample to show that the statement is false.

7. Figure *LMNO* is a quadrilateral. What additional information is necessary to prove that it is a trapezoid?

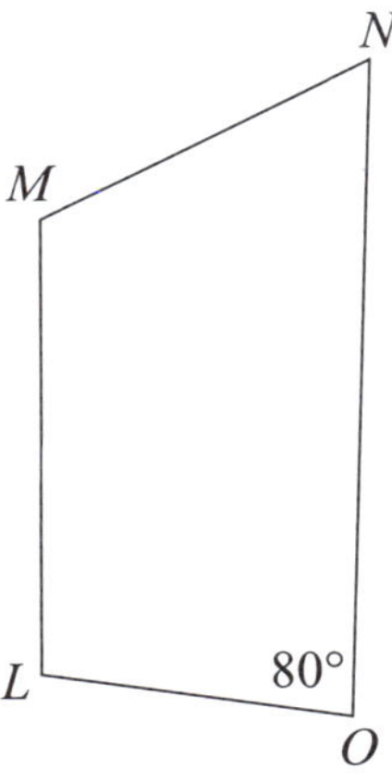

A. $m\angle LMN = 120°$

B. $m\angle ONM = 60°$

C. $\overline{LM} \parallel \overline{NO}$

D. $m\angle MLO = 110°$

8. If *AC* = 12 and *BC* = 2*AB*, what are the shortest possible lengths of *AB* and *BC*? (Use whole numbers only.)

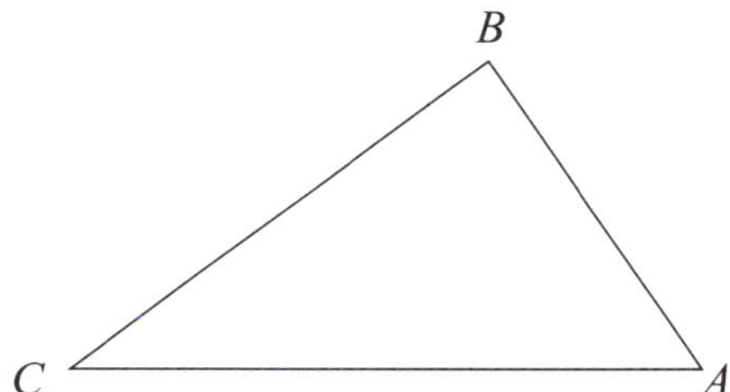

9. Find the ratio of the corresponding sides in each pair of similar figures.

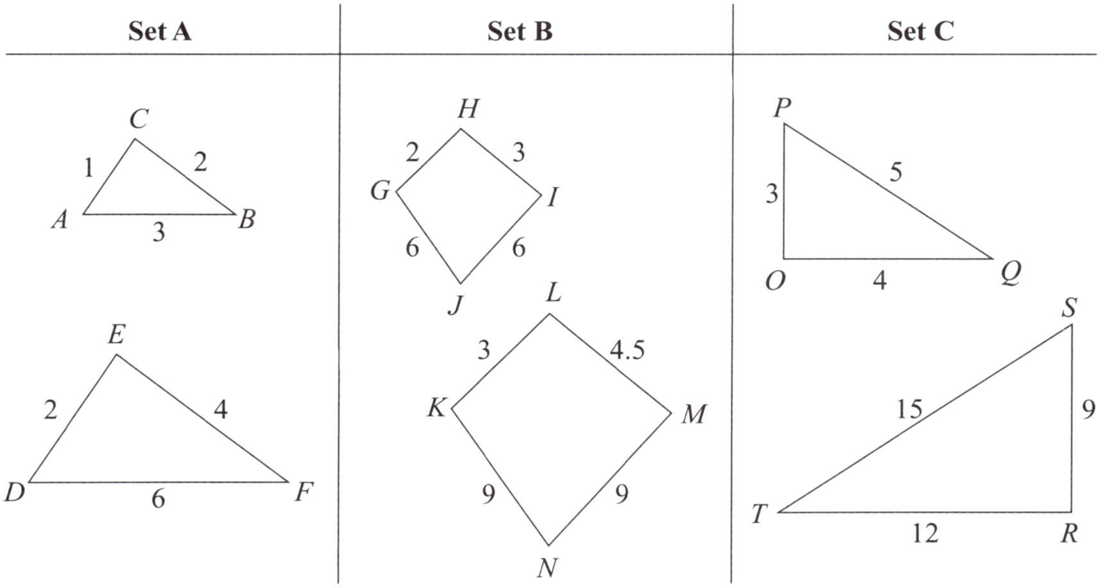

Set A	Set B	Set C

10. Use a two-column method to prove the following:

Given: $\overline{BD}$ **bisects** $\angle ABC$ **and**
$\overline{BD}$ **is a** $\perp$ **bisector of** $\overline{AC}$

Prove: $\triangle ABD \cong \triangle CBD$

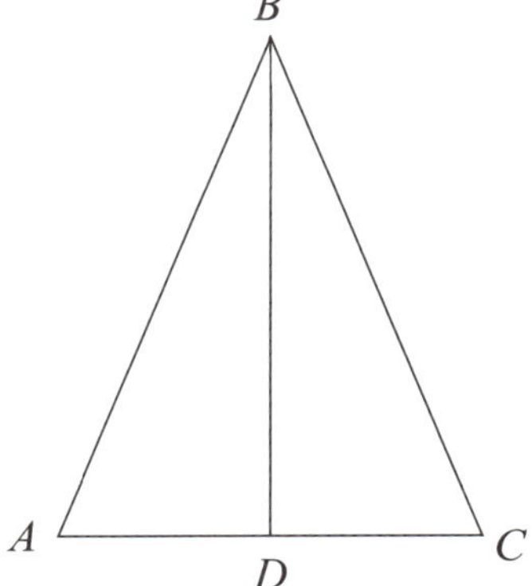

Answers to the Chapter Quiz

1. D

Both a parallelogram and supplementary angles have a definition using previously known words or terms. A plane is one of the undefined words of geometry.

2. Postulate.

It is a true statement that does not require a proof.

3. Max's answer is not reasonable.

$C = 2\pi r$

$\frac{8\pi}{2\pi} = \frac{2\pi r}{2\pi}$

$4 = r$

Max's answer is the diameter of the circle.

4. No

Stating that $\angle DEF$ must be an obtuse angle is not reasonable because complementary angles sum to 90° which is less than an obtuse angle. If $m\angle ABC = 30°$, then $m\angle DEF = 60°$.

5. 25.13 ft

Drawing a diagram is a useful strategy to solve this problem.

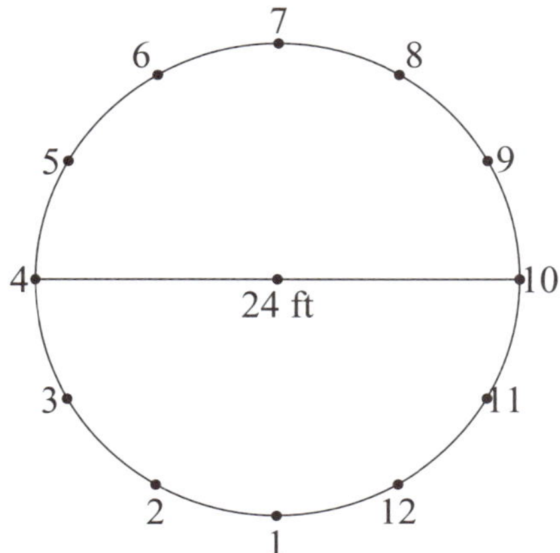

To find how far Jonah will travel from the location of seat 1 to seat 5, solve for the arc length. Since there are 12 equally spaced seats, the measure of each seat is $\frac{360°}{12} = 30°$. Jonah will move 4 spots, which is 120°. Arc length = $\frac{120°}{360°}(2\pi(12)) = \frac{1}{3}(24\pi) = 8\pi \approx 25.13 \text{ ft}$.

6. Not all rectangles are squares. A quadrilateral is a rectangle by having 2 pairs of congruent parallel sides and 4 right angles. It does not have to be a square, because all 4 sides do not have to be congruent. As a counterexample, let the length equal 6 and the width equal 4.

7. C

A trapezoid is a quadrilateral with one pair of parallel sides. Therefore, to prove figure *LMNO* is a trapezoid, $\overline{LM}$ must be parallel to $\overline{NO}$.

8. 5 and 10

Guess and check is a good strategy to find the length of AB and BC. Remember, the sum of two sides of a triangle must be greater than the third side.

AC	AB	$BC = 2AB$	
12	1	2	$1 + 2 \ngtr 12$
12	2	4	$2 + 4 \ngtr 12$
12	3	6	$3 + 6 \ngtr 12$
12	4	8	$4 + 8 \ngtr 12$
12	5	10	$5 + 10 > 12$

The shortest possible lengths of AB and BC are 5 and 10.

9. Conjecture: The ratios of corresponding side lengths in similar figures are equal.

Set A	Set B	Set C
$\frac{AB}{DE} = \frac{1}{2}$	$\frac{GH}{KL} = \frac{2}{3}$	$\frac{OP}{RS} = \frac{3}{9} = \frac{1}{3}$
$\frac{BC}{EF} = \frac{2}{4} = \frac{1}{2}$	$\frac{HI}{LM} = \frac{3}{4.5} = \frac{2}{3}$	$\frac{PQ}{ST} = \frac{5}{15} = \frac{1}{3}$
$\frac{AC}{DF} = \frac{3}{6} = \frac{1}{2}$	$\frac{IJ}{MN} = \frac{6}{9} = \frac{2}{3}$	$\frac{QO}{TR} = \frac{4}{12} = \frac{1}{3}$
	$\frac{JG}{NK} = \frac{6}{9} = \frac{2}{3}$	

10.

Statement	Reason
1. $\overline{BD}$ bisects $\angle ABC$ $\overline{BD}$ is a perpendicular bisector of AC	1. Given
2. $\angle ABD \cong \angle CBD$	2. Definition of angle bisector
3. $\angle ADB$ and $\angle CDB$ are right angles	3. Definition of perpendicular
4. $m\angle ADB = 90°$ $m\angle CDB = 90°$	4. Definition of right angles
5. $\angle ADB \cong \angle CDB$	5. Substitution
6. $\overline{BD} \cong \overline{BD}$	6. Reflexive Property
7. $\triangle ABD \cong \triangle CBD$	7. Angle-Side-Angle Postulate

Chapter 8

Logic

Your Goals for Chapter 8

1. You should be able to determine the converse, inverse, and contrapositive of a statement.

2. You should be able to determine if two propositions are logically equivalent.

3. You should be able to use direct and indirect proof to determine the validity of a short proof.

Standards

The following standards are assessed on Florida's Geometry End-of-Course exam either directly or indirectly:

MA.912.D.6.2: (Moderate) Find the converse, inverse, and contrapositive of a statement

MA.912.D.6.3: (Moderate) Determine whether two propositions are logically equivalent.

MA.912.D.6.4: (Moderate) Use methods of direct and indirect proof and determine whether a short proof is logically valid.

Conditional Statements

A **conditional statement** consists of two parts, the "if" clause contains the hypothesis and the "then" clause contains the conclusion.

Given an if-then statement, "if p, then q," three related statements can be created:

conditional statement	If p, then q.
converse	If q, then p.
inverse	If not p, then not q.
contrapositive	If not q, then not p.

Example:

Given the conditional statement:

If an angle is a right angle, then the angle measures 90°.

Identify the hypothesis and conclusion.

Write the converse, inverse and contrapositive and determine if they are true or false.

The hypothesis is contained in the "if" clause, so the hypothesis is: "an angle is a right angle." The conclusion is contained in the "then" clause, so the conclusion is: the angle measures 90°.

Converse: If an angle measures 90°, then the angle is a right angle. (*True*)

Inverse: If an angle is not a right angle, then the angle does not measure 90°. (*True*)

Contrapositive: If an angle does not measure 90°, then the angle is not a right angle. (*True*)

Example:

Given the conditional statement:

If a figure is a rectangle, then it has four sides.

Write the converse, inverse, and contrapositive and determine if they are true or false.

Converse: If a figure has four sides, then it is a rectangle. (False)

Inverse: If a figure is not a rectangle, then it does not have four sides. (False)

Contrapositive: If a figure does not have four sides, then it is not a rectangle. (True)

Truth Tables

Two statements can be proven logically equivalent with the construction of a truth table. A **truth table** gives the truth values of a compound sentence for all combinations of truth values of the sentences components.

In symbolic logic, as soon in truth tables, the letters p and q are used to represent the statements and symbols are used to represent the connectives.

Notation

not	$\sim$	negation
and	$\wedge$	conjunction
or	$\vee$	disjunction
if . . . then	$\rightarrow$	conditional
. . . if and only if	$\leftrightarrow$	biconditional

The truth value of a compound statement is determined from the truth value of its components.

For all truth tables of p and q, the first two columns must appear as:

p	q
T	T
T	F
F	T
F	F

Example:

Determine the truth values of the proposition $p \vee q$.

p	q	$p \vee q$
T	T	T
T	F	T
F	T	T
F	F	F

If p and q are statements, then the statement $p \vee q$ is true except when both p and q have truth values that are false.

Example:

Determine the truth values of the proposition $p \wedge q$.

p	q	$p \wedge q$
T	T	T
T	F	F
F	T	F
F	F	F

If p and q are statements, then the statement $p \wedge q$ is false except when both p and q have truth values that are true.

Example:

Determine whether $\sim(p \vee q)$ and $(\sim p \wedge \sim q)$ are logically equivalent.

Logical equivalence can be proven if all possible truth values for each proposition are equivalent.

p	q	$\sim p$	$\sim q$	$p \vee q$	$\sim(p \vee q)$	$(\sim p \wedge \sim q)$
T	T	F	F	T	F	F
T	F	F	T	T	F	F
F	T	T	F	T	F	F
F	F	T	T	F	T	T

Since the possible combinations of truth values for each proposition are the same, the propositions are logically equivalent.

Direct and Indirect Proof

A **direct proof** is an argument that establishes the truth of a conditional statement using direct use of the hypotheses.

An **indirect proof** is a type of proof in which a conditional statement is assumed false. The assumption leads to an impossibility, thus proving the statement true.

Example:

Prove the following statement using direct proof.

If n is odd, then $n^2 - 1$ is even.

To use a direct proof, assume the hypotheses as the starting point of the argument. If n is odd, then $n = 2k + 1$. The expression $2k + 1$ is odd because two times any integer, k, is even and one more than any even number is an odd number.

Since $n = 2k + 1$, $n^2 - 1 = (4k^2 + 4k + 1) - 1 = 4k^2 + 4k$.

Factoring 2 from the expression $4k^2 + 4k$, results in $2(2k^2 + 2k)$.

Since 2 is a factor of the expression $4k^2 + 4k$, it must be an even number. Therefore, $n^2 - 1$ is even. A good way to quickly check the accuracy of the conclusion is to select two odd numbers, say 7 and 13. If $n = 7$, then $n^2 - 1 = 7^2 - 1 = 49 - 1 = 48$. If $n = 13$, the $n^2 - 1 = 13^2 - 1 = 169 - 1 = 168$. Both 48 and 168 are even.

Example:

Prove the following statement using indirect proof.

If polygon XYZ is a triangle, then it has at most one right angle.

To use indirect proof, first assume that the statement is false. Therefore, assume $\triangle XYZ$ has more than one right angle. If more than one angle is a right angle, then $m\angle X = 90°$ and $m\angle Y = 90°$. Since the sum of the measures of the angles of a triangle equal 180°,

$m\angle X + m\angle Y + m\angle Z = 180°$

Substitution gives $90° + 90° + m\angle Z = 180°$.

Solving gives $m\angle Z = 0°$, which is not possible.

Therefore, there is no $\triangle XYZ$ that contradicts the given statement.

So the assumption that $m\angle X = 90°$ and $m\angle Y = 90°$ must be false.

Therefore, $\triangle XYZ$ has at most one right angle.

End-of-Chapter Quiz

For questions 1–3, use the conditional statement below:

If an angle measures 70°, then the angle is acute.

1. **Write the converse of the given conditional statement.**

2. **Write the inverse of the given conditional statement.**

3. **Write the contrapositive of the given conditional statement.**

4. **If a polygon is a parallelogram, then the polygon has 4 sides is a true conditional statement. Which of the following related statements is also true?**

   A. Converse

   B. Inverse

   C. Contrapositive

   D. Both inverse and contrapositive

5. Create a truth table for $p \vee \sim q$.

6. Determine if $\sim(p \wedge q)$ and $\sim p \vee \sim q$ are logically equivalent.

7. Determine if $p \wedge q$ and $\sim(p \vee q)$ are logically equivalent.

8. $\triangle XYZ$ is not isosceles. Using an indirect proof, prove that if altitude $\overline{XZ}$ is drawn, then Z is not the midpoint of $\overline{WY}$.

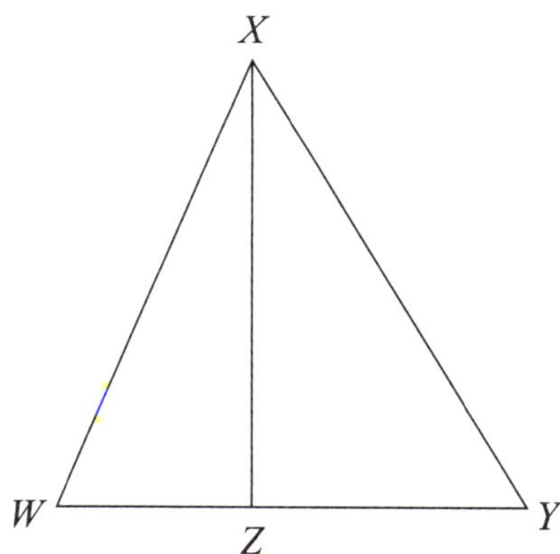

9. Using an indirect proof, prove that if $\triangle LMN$ is an isosceles triangle with base $\overline{LN}$ and segment $\overline{MO}$ is a median, then MO is an angle bisector.

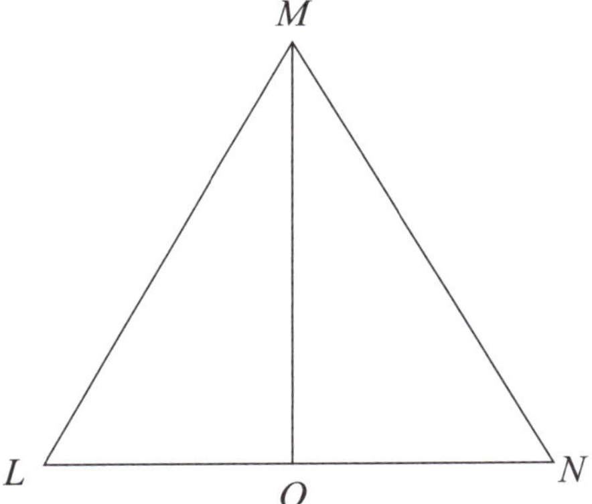

10. Using a direct proof, prove that if integers x and y are even, then their sum is even.

Answers to the Chapter Quiz

1. A converse switches the hypothesis and conclusion of the original conditional statement.

If an angle is acute, then the angle measures 70°.

2. The inverse negates the hypothesis and conclusion of the original conditional statement.

If an angle does not measure 70°, then the angle is not acute.

3. The contrapositive both negates and switches the hypothesis and conclusion of the original conditional statement.

If an angle is not acute, then the angle does not measure 70°.

4. C

Converse: If a polygon has 4 sides, then it is a parallelogram.

Inverse: If a polygon is not a parallelogram, then it does not have four sides.

Contrapositive: If a polygon does not have 4 sides, then it is not a parallelogram.

The converse is false because there are many 4 sided polygons that are not parallelograms.

The inverse is false because a polygon can have 4 sides without being a parallelogram.

The contrapositive is true because a polygon cannot be a parallelogram unless it has 4 sides.

5. Truth table for $p \vee \sim q$

p	q	$\sim q$	$p \vee \sim q$
T	T	F	T
T	F	T	T
F	T	F	F
F	F	T	T

6. A truth table can be used to determine if $\sim(p \wedge q)$ and $\sim p \vee \sim q$ are logically equivalent. $\vee$

p	q	$\sim p$	$\sim q$	$p \wedge q$	$\sim(p \wedge q)$	$\sim p \vee \sim q$
T	T	F	F	T	F	F
T	F	F	T	F	T	T
F	T	T	F	F	T	T
F	F	T	T	F	T	T

Since the possible combinations of truth values for each proposition are the same, the propositions are logically equivalent.

7. A truth table can be used to determine if $p \wedge q$ and $\sim(p \vee q)$ are logically equivalent.

p	q	$p \wedge q$	$p \vee q$	$\sim(p \vee q)$
T	T	T	T	F
T	F	F	T	F
F	T	F	T	F
F	F	F	F	T

Since not all the possible combinations of truth values for each proposition are the same, the propositions are not logically equivalent.

8.

To use indirect proof, assume that the conclusion is not true. Therefore, assume that Z is the midpoint of $\overline{WY}$. By definition of midpoint, $\overline{WZ} \cong \overline{ZY}$. Since $\overline{XZ}$ is an altitude, it is perpendicular to the opposite side of the vertex from which it is drawn. Therefore, $\overline{XZ} \perp \overline{WY}$. By definition of perpendicular, $\angle WZX$ and $\angle YZX$ are right angles and congruent because all right angles measure 90°. The reflexive property states that $\overline{XZ} \cong \overline{XZ}$, which results in $\triangle WXZ \cong \triangle YXZ$ by the Side-Angle-Side Postulate. Since corresponding parts of congruent triangles are congruent, $\overline{WX} \cong \overline{YX}$. Therefore, $\triangle WXY$ is proven to be an isosceles triangle. This contradicts the given statement, so that Z is not the midpoint of $\overline{WY}$.

9.

Since $\triangle LMN$ is isosceles, $\overline{LM} \cong \overline{MN}$ and $\angle MLO \cong \angle MNO$. To use indirect proof, assume that $\overline{MO}$ is not an angle bisector, so $\angle LMO$ is not congruent to $\angle NMO$. Since $\overline{MO}$ is a median, $\overline{LO} \cong \overline{NO}$. $\triangle MLO \cong \triangle MNO$ by the Side-Angle-Side Postulate. Since corresponding parts of congruent triangles are congruent, $\angle LMO \cong \angle NMO$, which contradicts the assumption that $\angle LMO$ is not congruent to $\angle NMO$. Therefore, $\overline{MO}$ is an angle bisector.

10. Since x and y are even, and $x = 2m$ and $y = 2n$. The sum $x + y = 2m + 2n$. Factoring a 2, $2m + 2n = 2(m + n)$. Since 2 is a factor of $x + y$, the sum is even.

Chapter 9

Circles

Your Goals for Chapter 9

1. You should be able to identify the parts of a circle.
2. You should be able to determine the measure of arcs and the angles created in a circle.
3. You should be able to solve problems using circumference, arc length, and areas of circles and sectors.
4. You should be able identify the center and radius of a circle and graph the circle on a coordinate plane from an equation in center-radius form.
5. You should be able to write an equation of a circle given the center and radius.

Standards

The following standards are assessed on Florida's Geometry End-of-Course exam either directly or indirectly:

MA.912.G.6.2: (Low) Define and identify: circumference, radius, diameter, arc, arc length, chord, secant, tangent and concentric circles.

MA.912.G.6.4: (Moderate) Determine and use measures of arcs and related angles (central, inscribed, and intersections of chords, secants and tangents).

MA.912.G.6.5: (High) Solve real-world problems using measures of circumference, arc length, and areas of circles and sectors.

MA.912.G.6.6: (Moderate) Given the center and the radius, find the equation of a circle in the coordinate plane or given the equation of a circle in center-radius form, state the center and the radius of the circle.

MA.912.G.6.7: (Moderate) Given the equation of a circle in center-radius form or given the center and the radius of a circle, sketch the graph of the circle.

Lines and Segments

A **radius** is a segment whose endpoints are the center and any point on the circle.

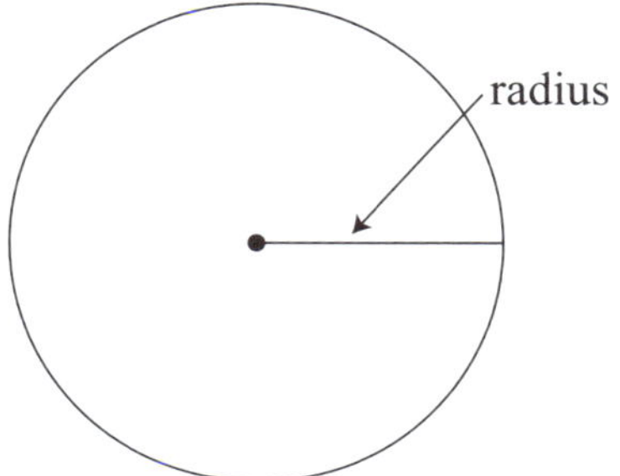

A **chord** is a segment whose endpoints are on the circle.

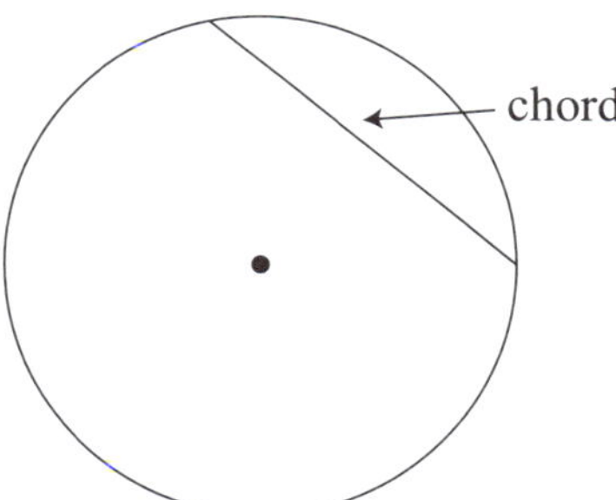

A **diameter** is a segment whose endpoints are on the circle and passes through the center. A diameter is the longest chord of the circle.

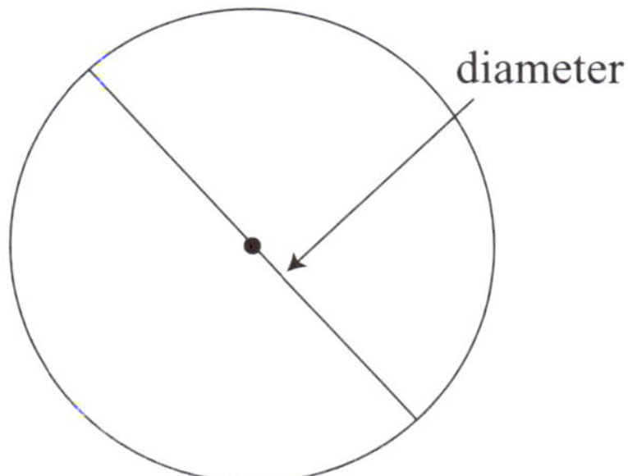

A **secant** is a line that intersects the circle at two points.

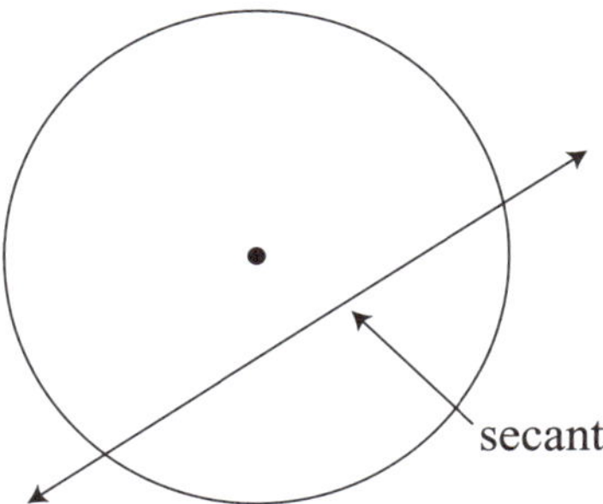

A **tangent** is a line that intersects the circle at exactly one point.

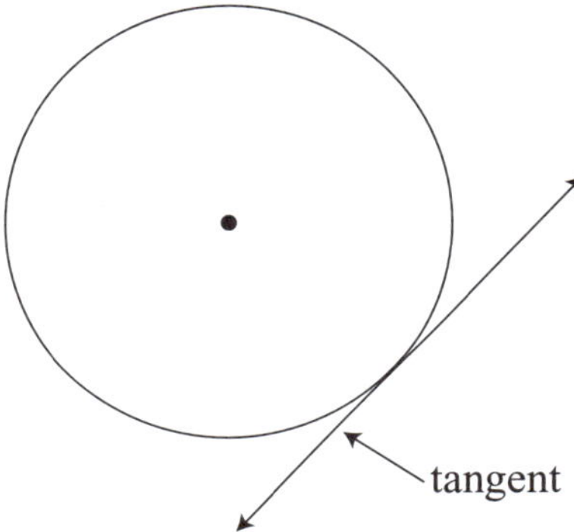

Example:

Identify each of the following in the given circle.

A. chord

B. secant

C. tangent

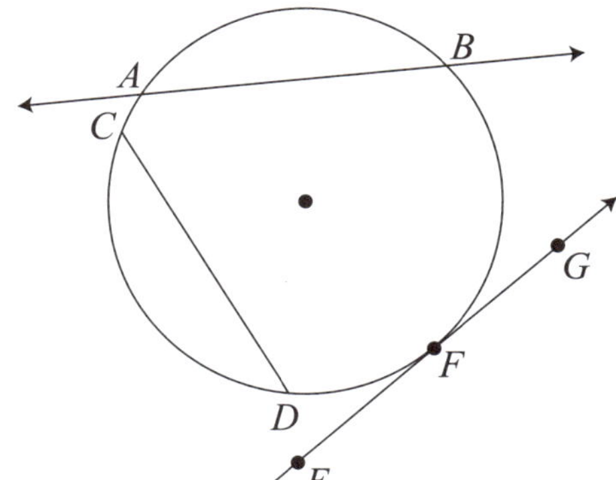

A. $\overline{CD}$ is a chord because it is a segment whose endpoints C and D are on the circle.

B. $\overleftrightarrow{AB}$ is a secant because it is a line that intersects the circle at two points, A and B.

C. $\overleftrightarrow{EG}$ is a tangent because it intersects the circle at exactly one point, F.

Arcs

An **arc** is a connected section of the circumference of a circle.

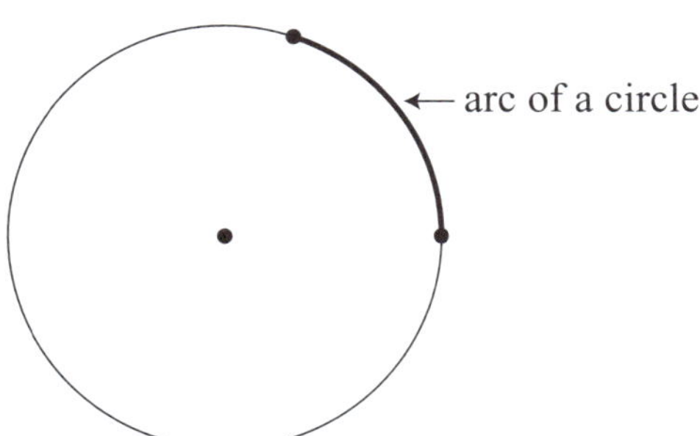

Types of Arcs

Minor arc measures less than 180°.

Major arc measures more than 180°.

Semicircle measures exactly 180°.

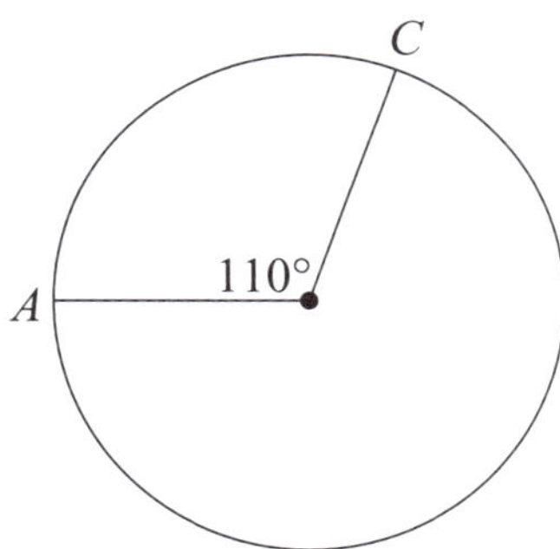

$\overset{\frown}{AC}$ is a minor arc and equals the measure of its central angle (110°).

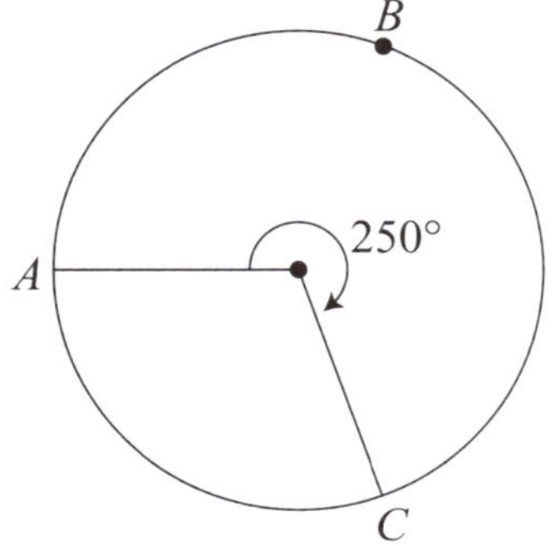

$\overset{\frown}{ABC}$ is a major arc and equals the measure of its central angle (250°).

Note: A major arc and a semicircle are always names using three letters.

Exercise 1

Identify each of the following in the circle to the right.

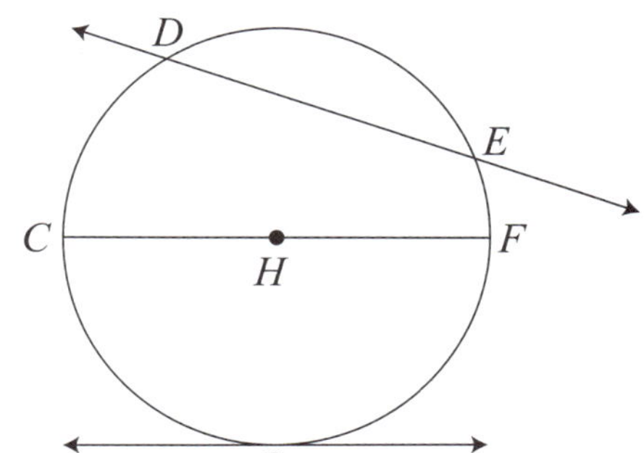

1. **secant**

2. **chord**

3. **minor arc**

4. **major arc**

Angle Types

A **central angle** of a circle is an angle whose vertex is located at the center of a circle.

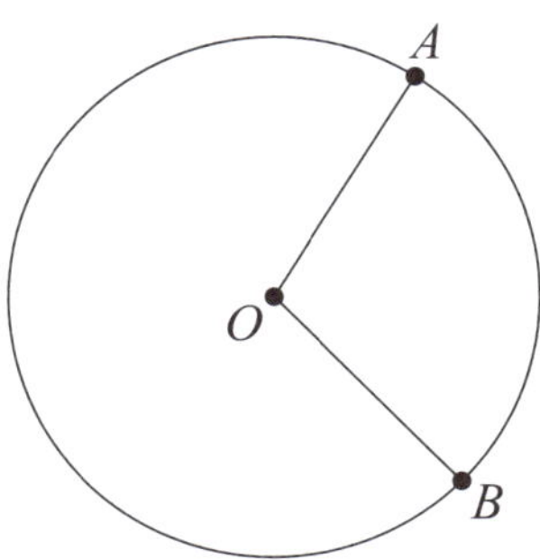

$\angle AOB$ is a central angle

An **inscribed angle** of a circle is an angle whose vertex is on a circle and whose sides each intersect the circle at another point.

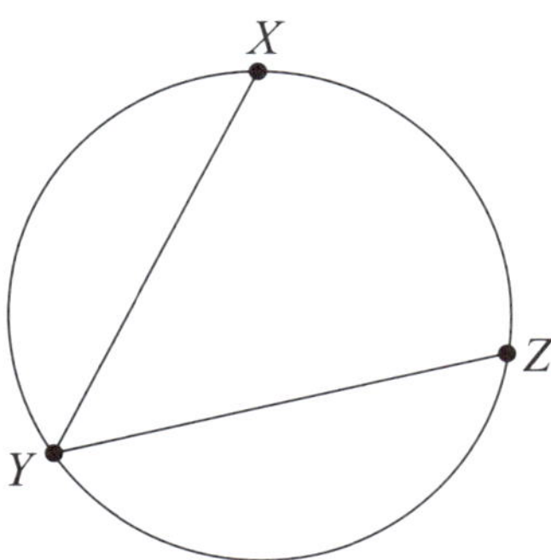

$\angle XYZ$ is an inscribed angle

Formulas

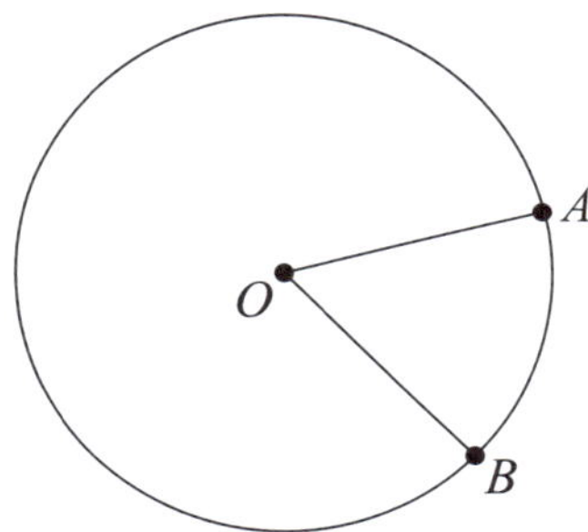

measure of a central angle = measure of its intercepted arc $m\angle AOB = m\overset{\frown}{AB}$

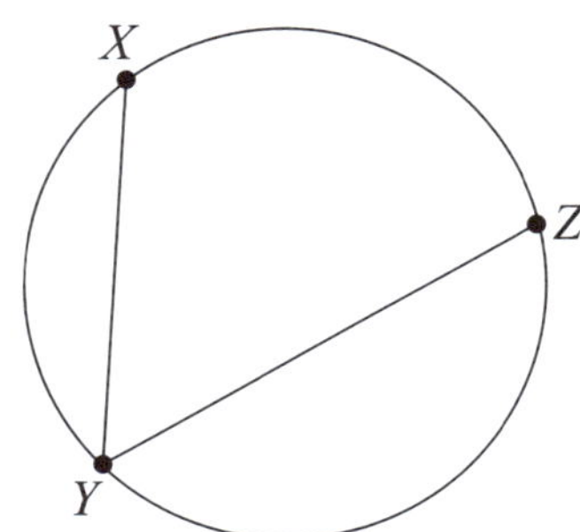

measure of an inscribed angle = ½ measure of its intercepted arc

$m\angle XYZ = \frac{1}{2}m\overset{\frown}{XZ}$

Example:

Given $m\widehat{LM} = 120°$ and $m\widehat{MN} = 100°$, find $m\angle LMN$.

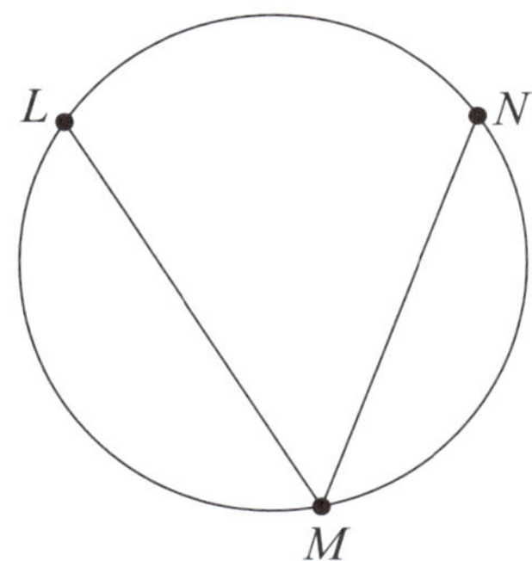

$\angle LMN$ is an inscribed angle, so

$$m\angle LMN = \frac{1}{2} m\widehat{LN}$$

Find mLN,

$$m\widehat{LN} + 120° + 100° = 360°$$

$$m\widehat{LN} + 220° = 360°$$

$$m\widehat{LN} = 140°$$

Since $m\angle LMN = \frac{1}{2} m\widehat{LN}$,

$$m\angle LMN = \frac{1}{2}(140°)$$

$$m\angle LMN = 70°$$

Example:

Given $m\widehat{GHE} = 210°$, find $m\angle EFG$.

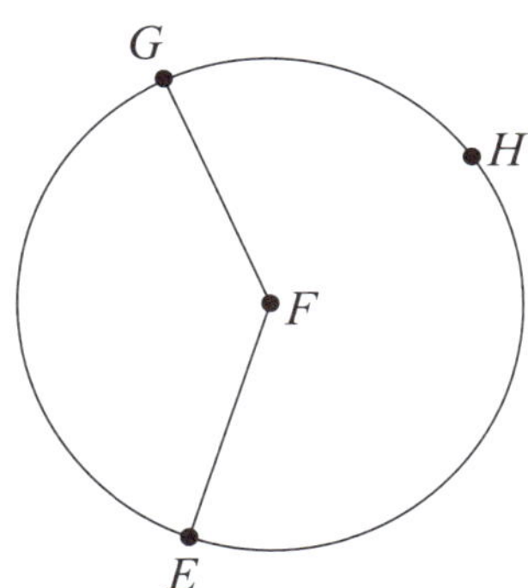

$\angle EFG$ is a central angle, so

$m\angle EFG = m\widehat{EG}$

Find $m\widehat{EG}$,

$m\widehat{EG} + 210° = 360°$

$m\widehat{EG} = 150°$

Since $m\angle EFG = m\widehat{EG}$

$m\angle EFG = 150°$

Note: If an inscribed angle intersects a semicircle, then the angle measure must be 90°, as shown in the following diagram.

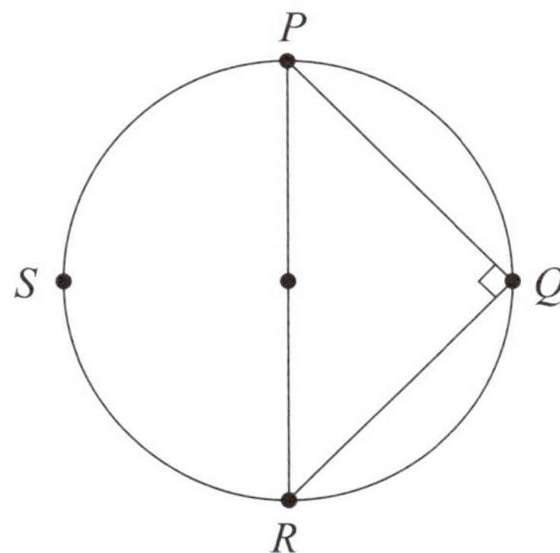

$\overline{PR}$ is a diameter, so $m\widehat{PQR} = 180°$. Then $m\angle PQR = \left(\frac{1}{2}\right)(180°) = 90°$.

Angles Formed by Secants and Tangents

The measure of an angle formed by two secants, two tangents, or a secant and a tangent drawn from a point outside the circle is equal to half of the difference of the intercepted arcs.

2 secants	2 tangents	secand and tangent
$m\angle DFE = \frac{1}{2}(m\widehat{ABC} - m\widehat{DE})$	$m\angle CDA = \frac{1}{2}(m\widehat{ABC} - m\widehat{AC})$	$m\angle CED = \frac{1}{2}(m\widehat{ABC} - m\widehat{CD})$

Example:

Given $m\widehat{MLP} = 200°$, find $m\angle NOP$.

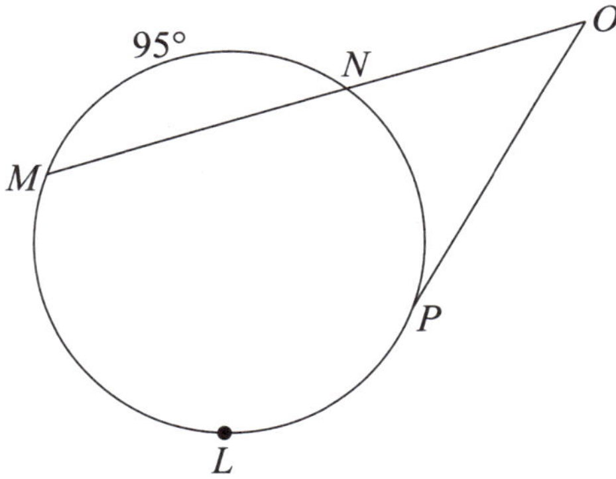

$\angle NOP$ is formed by a secant and a tangent, therefore,

$$m\angle NOP = \frac{1}{2}(m\widehat{MLP} - m\widehat{NP}).$$

$m\widehat{MLP} = 200°$

$m\widehat{NP} = 65°$ because the sum of the arcs must equal 360°. (360° – 200° – 95° = 65°)

$$m\angle NOP = \frac{1}{2}(200° - 65°)$$

$$m\angle NOP = \frac{1}{2}(135°)$$

$$m\angle NOP = 67.5°$$

Exercise 2

1. Find $m\angle DGE$

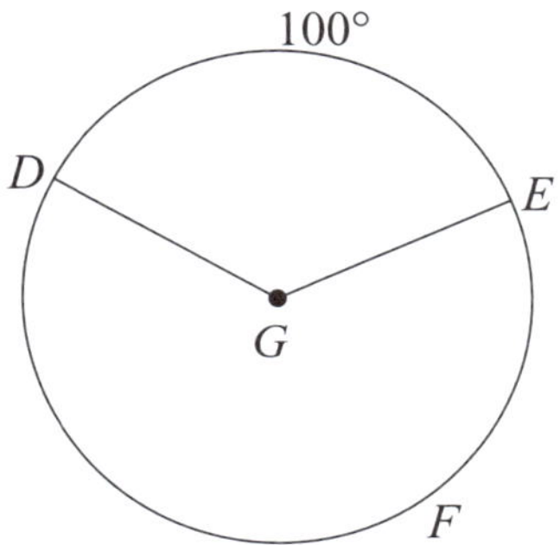

2. Find $m\angle PRQ$

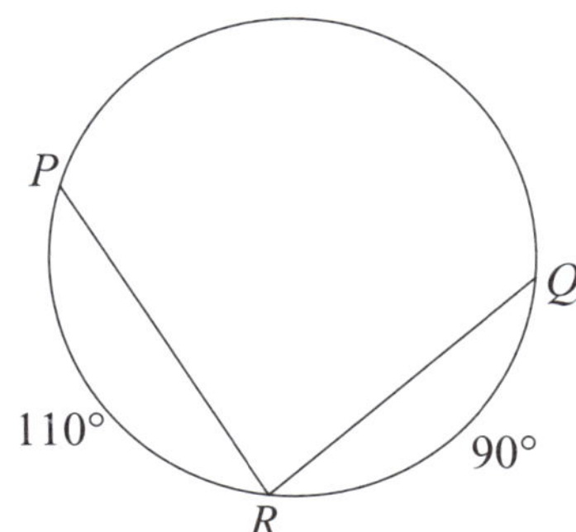

3. **Find $m\widehat{XY}$**

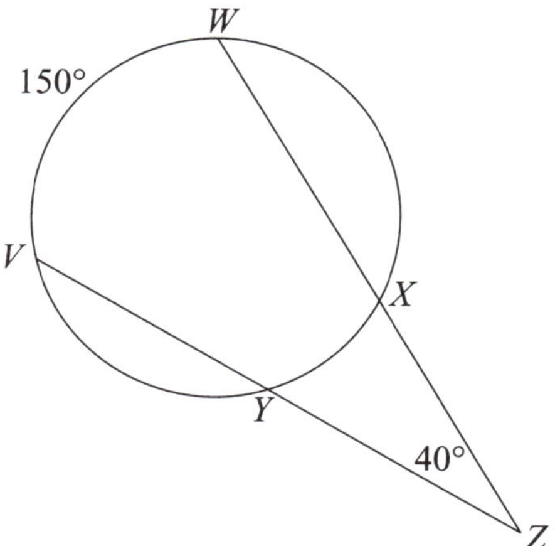

4. **Given $m\widehat{ADC} = 250°$, find $m\angle ABC$**

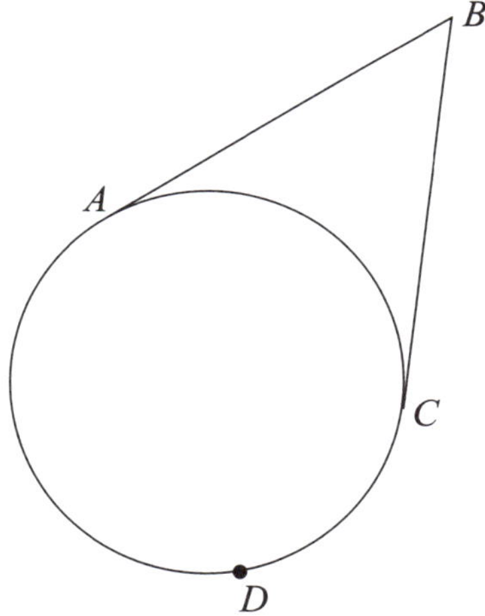

Circumference and Area

The **circumference of a circle** is the length around the outside of the circle. Circumference can be found using either of the formulas below:

$$C = 2\pi r \text{ or } \pi d$$

The **area of a circle** refers to the number of square units within the circle. Area of a circle can be found using the following formula.

$$A = \pi r^2$$

Note: π is a constant that is normally rounded off to 3.14.

Example:

Find the circumference and area of the given circle.

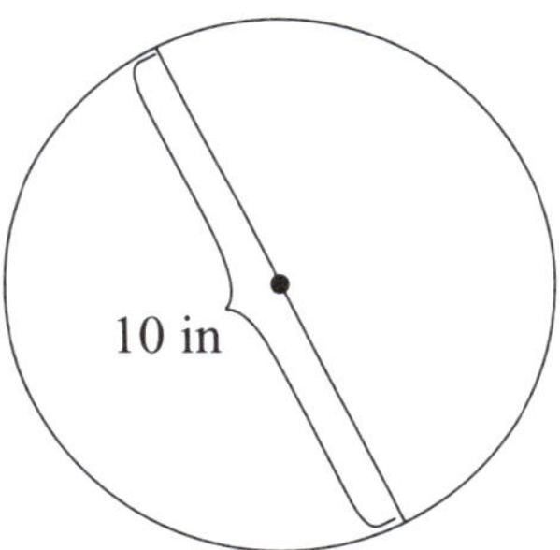

The diameter of the circle is 10 in. To find the circumference, substitute into the formula $C = \pi d$.

$C = \pi d$

$C = \pi(10 \text{ in})$

$C = 10\pi$ in.

The radius of the circle is 5 in because it is half the radius. To find the area, substitute into the formula $A = \pi r^2$.

$A = \pi(5 \text{ in.})^2$

$A = 25\pi$ square inches

Example:

If the area of a circle is approximately 28.26 cm^2, what is the length of the radius?

To find the length of the radius, substitute the area of the circle into the following formula:

$$28.26 = \pi r^2$$

$$28.26 = (3.14)r^2$$

$$\frac{28.26}{3.14} = \frac{(3.14)r^2}{3.14}$$

$$9 = r^2$$

$$\sqrt{9} = \sqrt{r^2}$$

$$3 = r$$

The radius of the circle is 3 cm.

Example:

The Super Soup Company is creating a new label for their soup cans. What is the length of the label needed to wrap around the can? Round off your answer to the nearest hundredth.

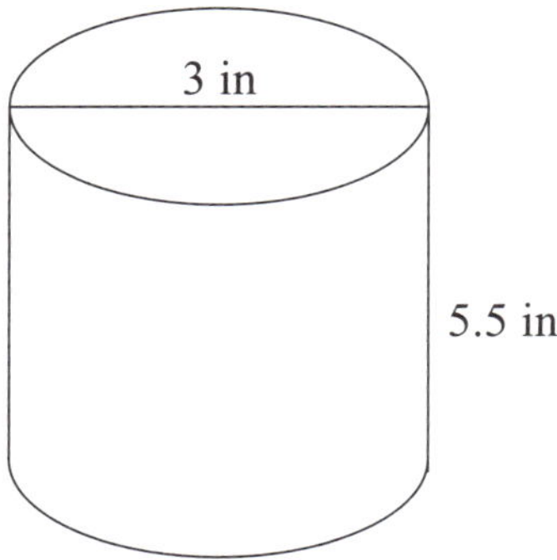

The circumference of the base of the can is the length around the outside of the can. Therefore, the length of the label is equal to the circumference of the can. The diameter of the base is 3in. Substitute the diameter into the following formula:

$C = \pi d$

$C = (3.14)(3 \text{ in})$

$C \approx 9.42 \text{ in}$

The length of the label on the soup can is approximately 9.42 inches.

Arc Length

The **length of an arc** can be found using the following formula:

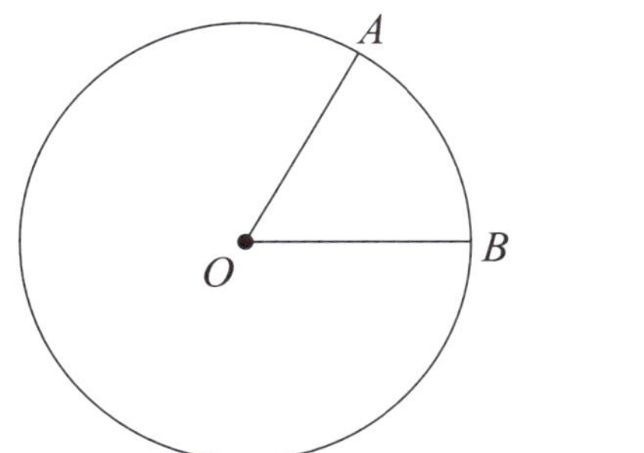

$$\text{length of } \overset{\frown}{AB} = \frac{m\angle AOB}{360°} \cdot 2\pi r$$

The length of an arc is a fraction of the circumference of the whole circle.

Note that earlier in this chapter, we showed how to determine the measure of an arc in degrees.

Example:

Find the length of $\overset{\frown}{CD}$.

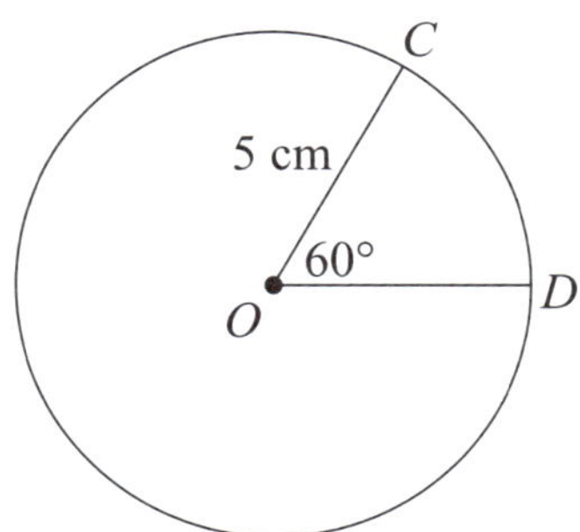

$$\text{length of } \overset{\frown}{CD} = \frac{m\angle COD}{360^\circ} \cdot 2\pi r$$

$$\text{length of } \overset{\frown}{CD} = \frac{60^\circ}{360^\circ} \cdot 2\pi(5)$$

$$\text{length of } \overset{\frown}{CD} = \frac{1}{6} \cdot 10\pi$$

$$\text{length of } \overset{\frown}{CD} = \frac{5}{3}\pi$$

The length of $\overset{\frown}{CD}$ is $\frac{5}{3}\pi$ or approximately 5.23 cm.

Example:

Allie is building a circular deck around the corner of her rectangular house. According to the picture, how much railing does she need to enclose her deck?

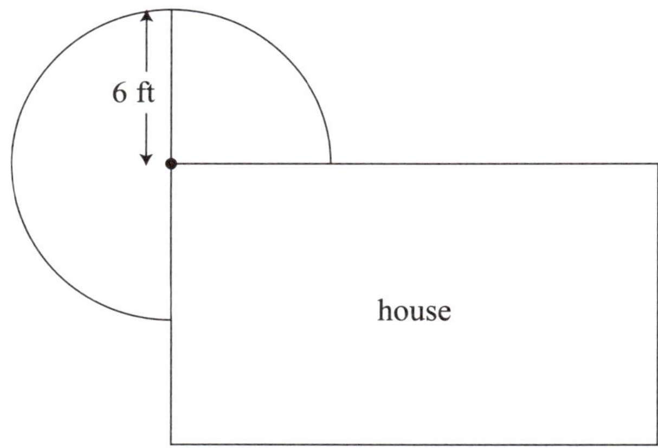

Since the house is rectangular, the corner of the house measures 90°. Therefore, the measure of the central angle related to the deck is 270°. The length of the railing for the deck is $\frac{270^\circ}{360^\circ}(2\pi(6)) = \frac{3}{4}(12\pi) = 9\pi \approx 28.26 \text{ ft}$, using $\pi = 3.14$.

Area of a Sector

A **sector** is a pie shaped region of a circle bounded by an arc and an angle.

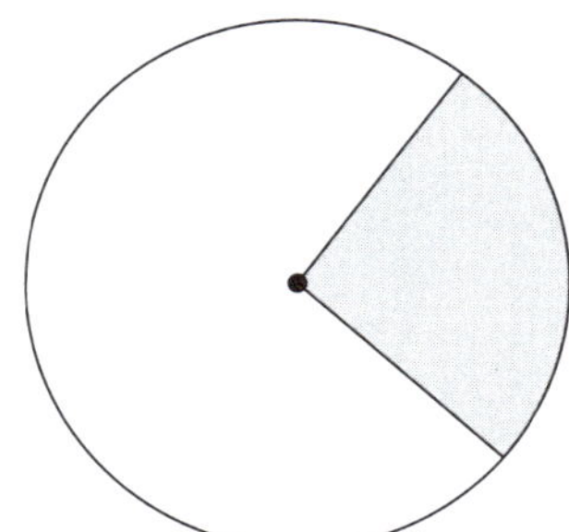

the shaded area is a sector

The **area of a sector** can be found using the following formula:

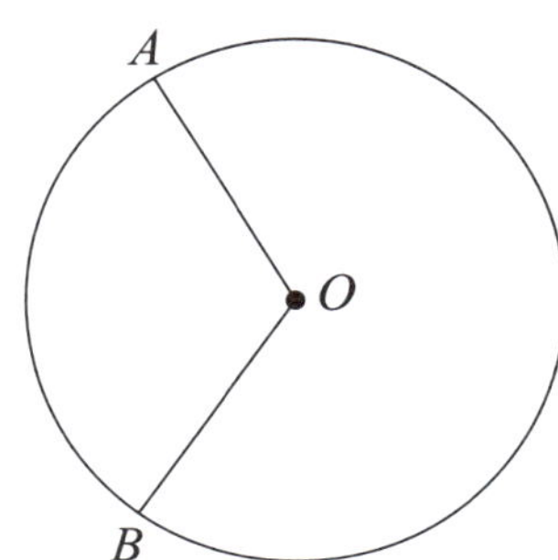

Area of sector $AOB = \frac{m\angle AOB}{360^o} \cdot \pi r^2$

The area of a sector is a fraction of the area of the whole circle.

Example:

Find the area of sector *COD*.

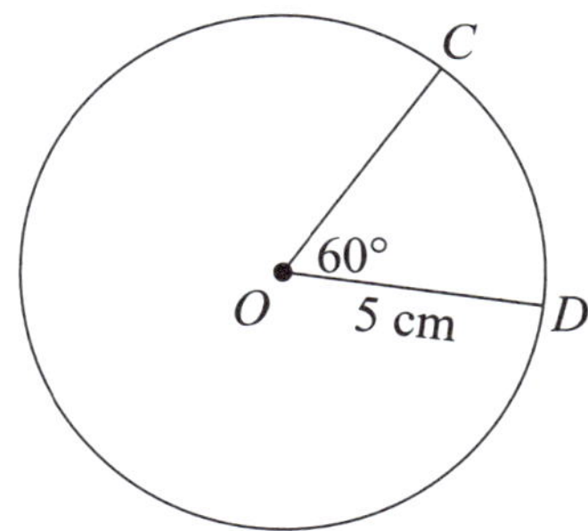

area of sector $COD = \frac{m\angle COD}{360°} \cdot \pi r^2$

$$\text{area of sector } COD = \frac{60°}{360°} \cdot \pi(5)^2$$

$$\text{area of sector } COD = \frac{1}{6} \cdot 25\pi$$

$$\text{area of sector } COD = \frac{25}{6}\pi \approx 13.08 \text{ cm}^2.$$

Example:

Patty is selling pie at the local bake sale. She has cut each of her 9-inch circular pies into eighths. What is the area of one slice of pie?

Each slice of the pie is a sector of the circular pie. The diameter of the pie is 9 inches, so the radius is 4.5 inches and a slice represents $\frac{1}{8}$ of the pie. Therefore, the area of a pie slice is equal to $\frac{1}{8}[\pi(4.5)^2] = \frac{1}{8}(20.25\pi) \approx 7.95$ square inches.

Exercise 3

1. Find the length of $\overset{\frown}{AB}$. Round off your answer to the nearest hundredth.

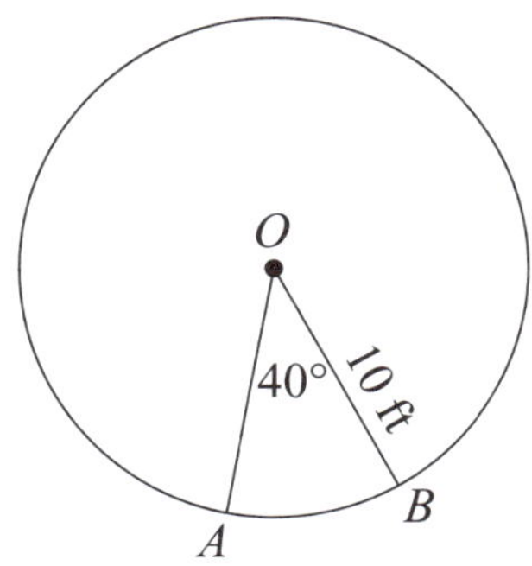

2. Find the area of sector *AOB*.

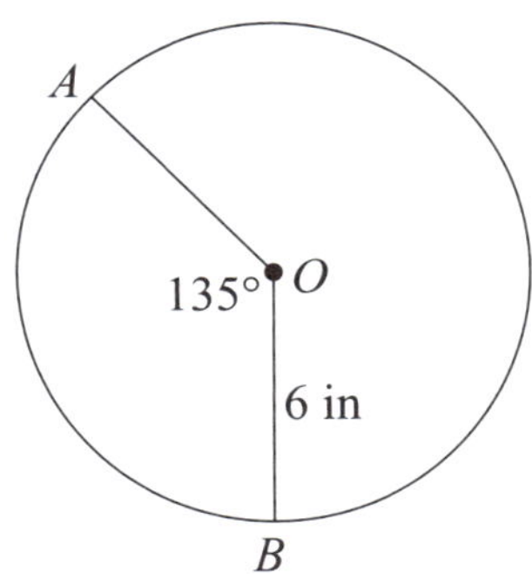

Equation of a Circle

The standard form for the equation of a circle is used to define a circle on the coordinate plane. The equation is as follows:

$$(x-h)^2+(y-k)^2=r^2$$

where h and k are the x- and y-coordinates of the center of the circle and r is the radius

Example:

Write the equation of a circle with center (–4, 0) and a diameter that measures 8 units.

Since the center is located at (–4, 0), $h = -4$ and $k = 0$. The radius is half the length of the diameter, so the radius measures 4 units. Substituting into the standard form of the equation of the circle,

$$(x-h)^2+(y-k)^2=r^2$$

$$(x-(-4))^2+(y-(0))^2=(4)^2$$

$$(x+4)^2+(y)^2=16$$

The equation of the circle is $(x+4)^2+y^2=16$.

Example:

Find the equation of the circle on the coordinate plane.

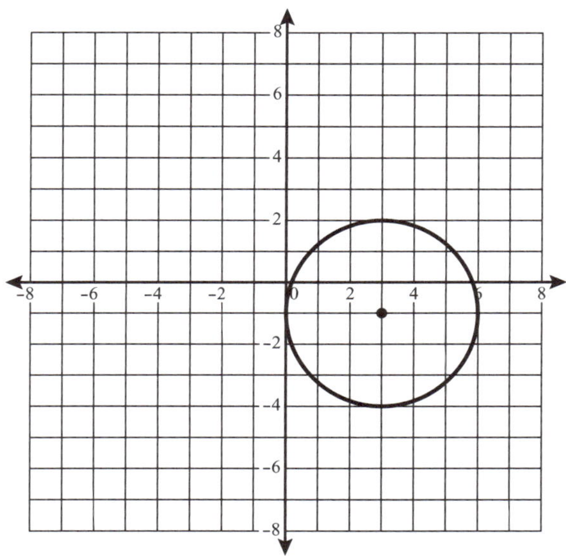

Determine the coordinates of the center.

$(h, k) = (3, -1)$

Determine the length of the radius.

$r = 3$ units

Substitute into the standard form of the equation of the circle.

$$(x-h)^2 + (y-k)^2 = r^2$$

$$(x-3)^2 + (y-(-1))^2 = (3)^2$$

$$(x-3)^2 + (y+1)^2 = 9$$

The equation of the circle is $(x-3)^2 + (y+1)^2 = 9$.

Example:

Determine the center and the radius of the circle.

$$(x+2)^2 + y^2 = 49$$

The center is located at the point (h, k) given the formula $(x-h)^2 + (y-k)^2 = r^2$.

Therefore,

$$(x+2)^2 + y^2 = 49$$

$$(x-(-2))^2 + (y-0)^2 = 49$$

$h = -2$ and $k = 0$

center = (–2, 0)

The radius is the value of r given the formula $(x-h)^2 + (y-k)^2 = r^2$.

Therefore,

$$r^2 = 49$$

$$\sqrt{r^2} = \sqrt{49}$$

$r = 7$ and -7

The radius is 7 because length has to be a positive value.

The center is (–2, 0) and the radius is 7.

Example:

Graph $x^2 + (y - 3)^2 = 25$

To graph the circle, determine the center and the radius.

Center: $(h, k) = (0, 3)$

Radius = $\sqrt{25} = 5$

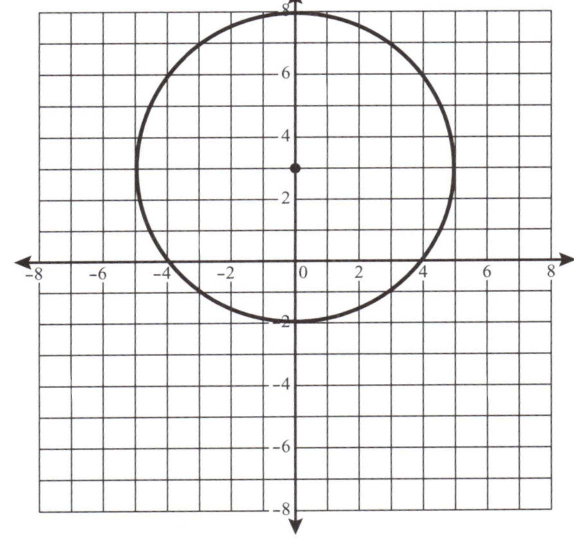

Exercise 4

1. Write the equation of a circle with center (–1, –5) and a radius that measures 12 units.

2. Determine the center and the radius of the circle.

$$x^2 + (y - 3)^2 = 20$$

3. Graph $(x + 2)^2 + (y - 1)^2 = 4$

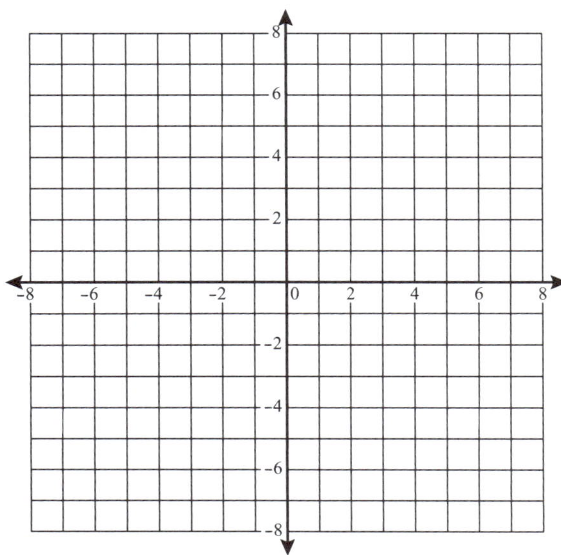

End-of-Chapter Quiz

Identify each of the following as a radius, chord, diameter, secant, or tangent.

1. $\overline{AE}$

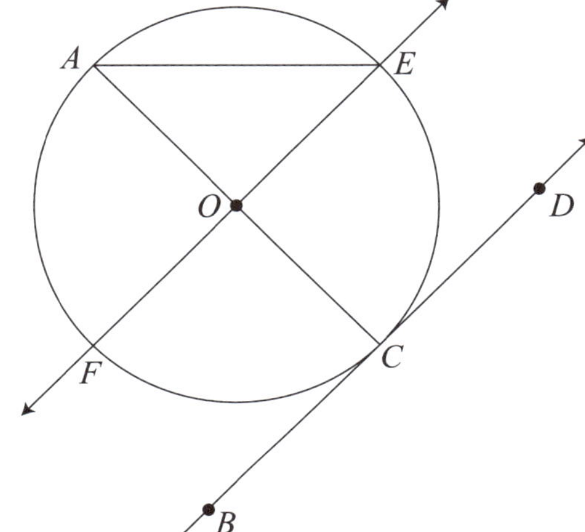

2. $\overline{AO}$

3. $\overline{AC}$

4. $\overline{EF}$

5. $\overleftrightarrow{EF}$

6. $\overleftrightarrow{BD}$

7. Find $m\widehat{BCD}$.

A. 80°

B. 140°

C. 160°

D. 180°

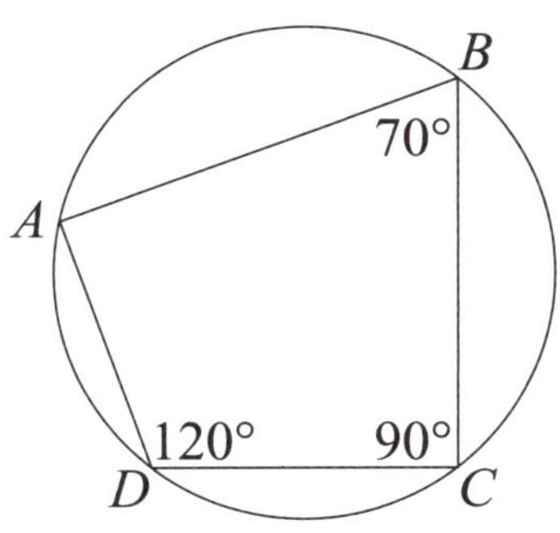

8. Find $m\angle LPM$.

A. 35°

B. 55°

C. 70°

D. 110°

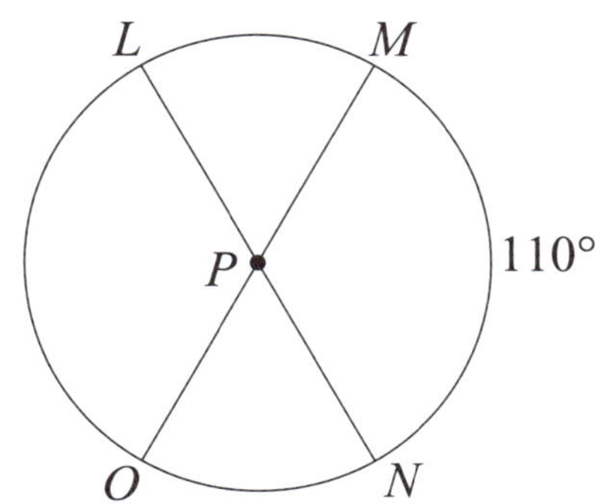

9. Find $m\overset{\frown}{RT}$.

A. 20°

B. 25°

C. 30°

D. 35°

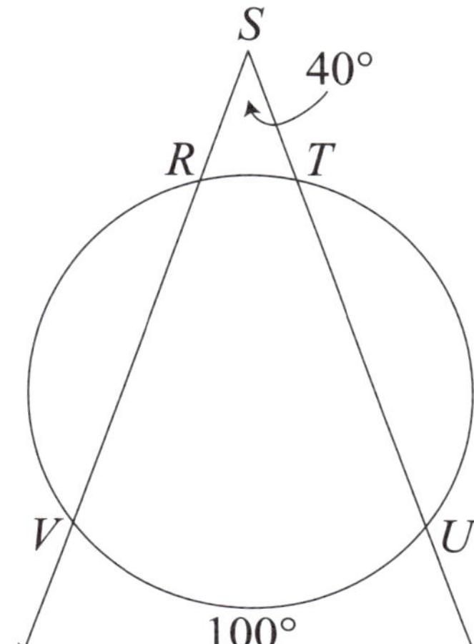

10. The area of a circle is approximately 153.86 cm^2. Find the circumference of the circle.

11. Chloe the Clown rides a unicycle. The diameter of the unicycle wheel is 20 inches. Approximately how many times will the wheel turn to ride 10 feet?

12. Margaret wants to plant grass in her backyard. The bags of seed at the store state that one bag covers 150 square feet. If the semicircle below represents Margaret's backyard, how many bags of seed will she need to buy?

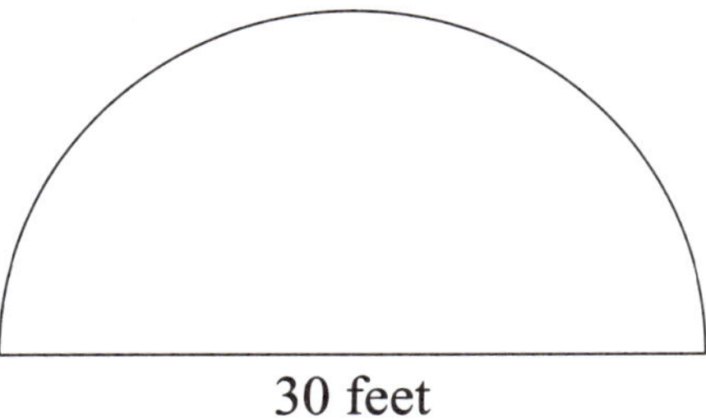

13. Find the length of $\overset{\frown}{ABC}$.

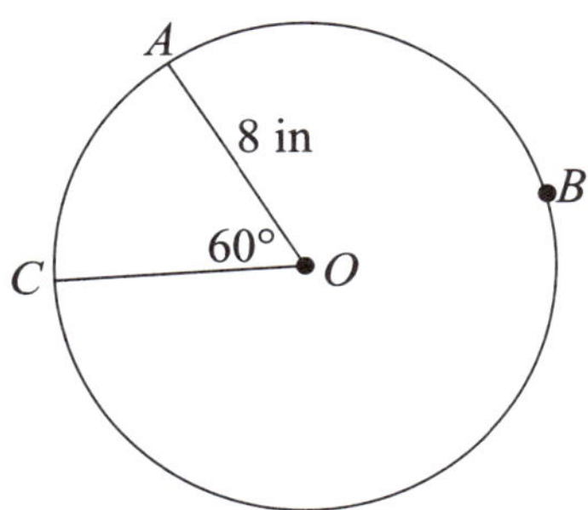

14. Find the area of sector *BOC*.

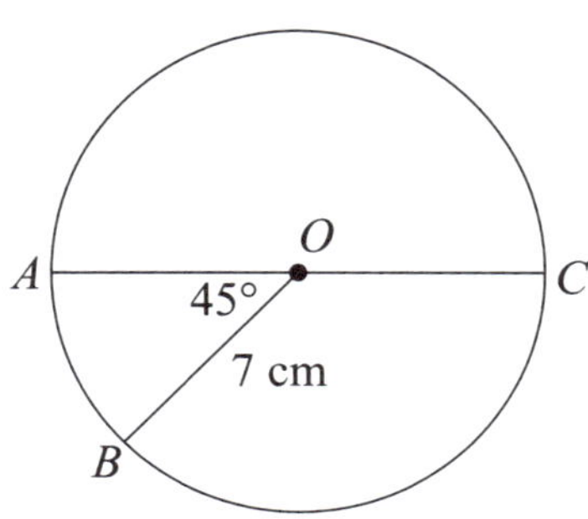

15. The figure shows 12 equally space points around the circle. What is the length of $\overset{\frown}{AE}$?

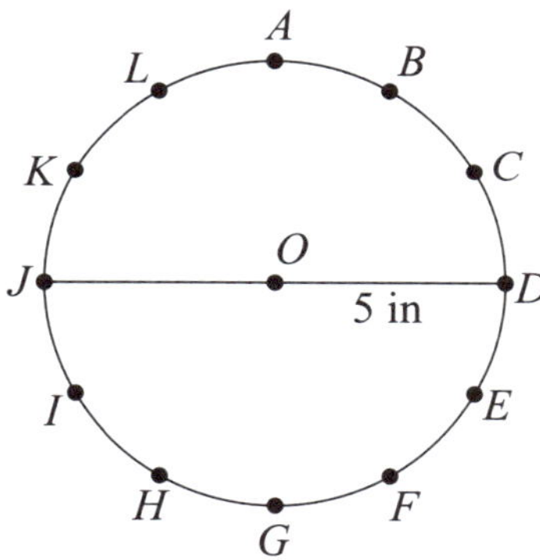

16. Anika's paper fan is a sector with radius 6 inches and a central angle of 160°. What is the area of Anika's fan?

17. Which of the following is the center of the circle represented by $(x - 1)^2 + (y + 1)^2 = 81$?

A. $(1, -1)$

B. $(-1, 1)$

C. $(1, 1)$

D. $(-1, -1)$

18. Which of the following is the equation of a circle with center at (–4, 0) and radius of 4?

A. $(x - 4)^2 + y^2 = 4$

B. $(x - 4)^2 + y^2 = 16$

C. $(x + 4)^2 + y^2 = 4$

D. $(x + 4)^2 + y^2 = 16$

19. Graph $(x + 3)^2 + (y - 2)^2 = 9$.

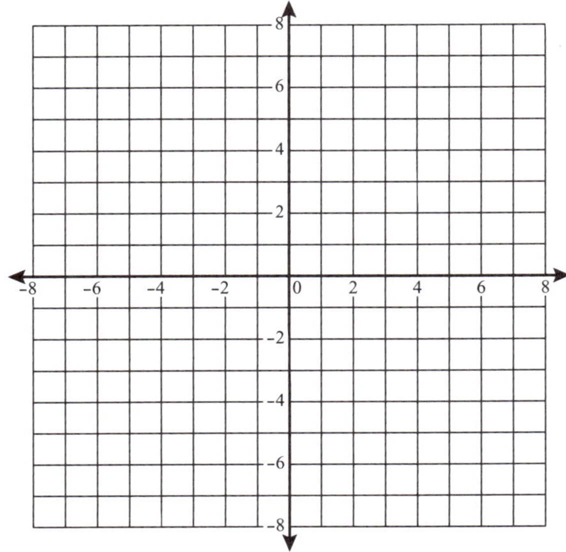

20. Find the area of a circle represented by $x^2 + y^2 = 64$.

Answers to Exercise 1

1. $\overleftrightarrow{DE}$

$\overleftrightarrow{DE}$ is a secant because it intersects the circle at 2 points.

2. $\overline{CF}$ or $\overline{DE}$

$\overline{CF}$ or $\overline{DE}$ can be a chord because their endpoints are on the circle.

3. $\overparen{CD}$, $\overparen{CE}$, $\overparen{DE}$, $\overparen{DF}$, $\overparen{EF}$, $\overparen{EG}$, $\overparen{FG}$, $\overparen{DG}$

$\overparen{CD}$, $\overparen{CE}$, $\overparen{DE}$, $\overparen{DF}$, $\overparen{EF}$, $\overparen{EG}$, $\overparen{FG}$, $\overparen{DG}$ are all considered minor arcs because their measures are less than 180°.

4. $\overparen{CDG}$, $\overparen{DFG}$, $\overparen{CGE}$

$\overparen{CDG}$, $\overparen{DFG}$, $\overparen{CGE}$ are all considered major arcs because their measures are greater than 180°.

Answers to Exercise 2

1. 100°

$m\angle DGE = 100°$ because a central angle is equal to its intercepted arc.

2. 80°

$m\widehat{PR} + m\widehat{QR} + m\widehat{PQ} = 360° \Rightarrow 110° + 90° + m\widehat{PQ} = 360° \Rightarrow 200° + m\widehat{PQ} = 360°$, so $m\widehat{PQ} = 160°$. $\angle PQR$ is an inscribed angle and is equal to half of its intercepted arc, $\widehat{PQ}$. Thus, $m\angle PQR = \left(\frac{1}{2}\right)(160°) = 80°$.

3. 70°

$$m\angle VZW = \frac{1}{2}(m\widehat{VW} - m\widehat{XY})$$

$$40° = \frac{1}{2}(150° - m\widehat{XY})$$

$$80° = 150° - m\widehat{XY}$$

$$70° = m\widehat{XY}$$

4. 70°

$$m\angle ABC = \frac{1}{2}(m\widehat{ADC} - m\widehat{AC})$$

$$m\widehat{AC} = 360° - 250° = 110°$$

$$m\angle ABC = \frac{1}{2}(250° - 110°)$$

$$m\angle ABC = \frac{1}{2}(140°)$$

$$m\angle ABC = 70°$$

Answers to Exercise 3

1. 6.98 ft

$$\text{length of } \overset{\frown}{AB} = \frac{m\angle AOB}{360°} \cdot 2\pi r$$

$$\text{length of } \overset{\frown}{AB} = \frac{40°}{360°} \cdot 2\pi(10)$$

$$\text{length of } \overset{\frown}{AB} = \frac{1}{9} \cdot 20\pi$$

$$\text{length of } \overset{\frown}{AB} \approx 6.98 \text{ ft}$$

2. 42.39 in^2

$$\text{area of sector } AOB = \frac{m\angle AOB}{360°} \cdot \pi r^2$$

$$\text{area of sector } AOB = \frac{135°}{360°} \cdot \pi(6)^2$$

$$\text{area of sector } AOB = \frac{3}{8} \cdot 36\pi$$

$$\text{area of sector } AOB \approx 42.39 \text{ in}^2$$

Answers to Exercise 4

1. $(x + 1)^2 + (y + 5)^2 = 144$

The standard form of the equation of a circle is $(x - h)^2 + (y - k)^2 = r^2$. (h, k) is the center and r is the length of the radius.

$$(x-(-1))^2+(y-(-5))^2=12^2$$

$$(x+1)^2+(y+5)^2=144$$

2. The center of the circle is (0, 3) and the radius = $\sqrt{20}$.

3.

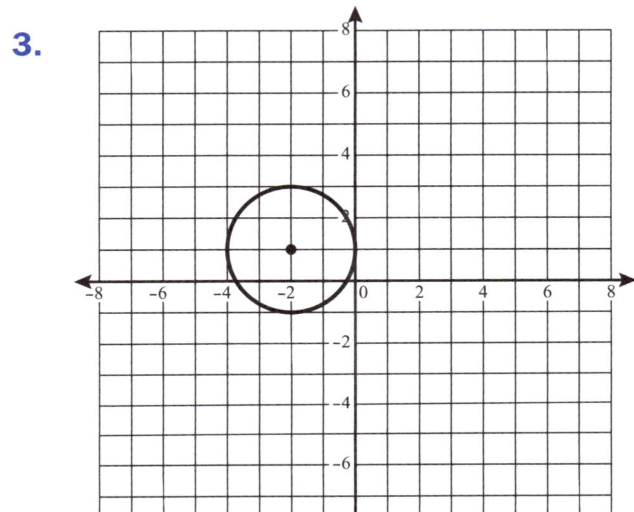

Note: the center is (–2, 1) and the radius = $\sqrt{4}$ = 2.

Answers to the Chapter Quiz

1. chord

$\overline{AE}$ is a chord because the endpoints are on the circle.

2. radius

$\overline{AO}$ is a radius because one endpoint is the center and the other endpoint is on the circle.

3. diameter and chord

$\overline{AC}$ is a diameter and a chord. All diameters are chords and $\overline{AC}$ is a diameter because both endpoints are on the circle and it passes through the center.

4. chord

$\overline{EF}$ is a diameter and a chord because it is a segment whose endpoints are on the circle. Also, it passes through the center.

5. secant

$\overleftrightarrow{EF}$ is a secant because it is a line that intersects the circle at 2 points.

6. tangent

$\overleftrightarrow{BD}$ is a tangent because it is a line that intersects the circle at one point.

7. C

Figure $ABCD$ is a quadrilateral. The sum of the angles of a quadrilateral is 360°, so $m\angle DAB = 360° - 70° - 90° - 120° = 80°$. $m\angle DAB = \frac{1}{2} m\widehat{BCD} \Rightarrow 160° = m\widehat{BCD}$.

8. C

$m\angle MPN = 110°$ because a central angle is equal to its intercepted arc. Since $\overline{LN}$ is a diameter, $m\angle LPM + m\angle MPN = 180° \Rightarrow m\angle LPM + 110° = 180°$, so $m\angle LPM = 70°$.

9. A

$$m\angle VSU = \frac{1}{2}(m\widehat{VU} - m\widehat{RT})$$

$$40° = \frac{1}{2}(100° - m\widehat{RT})$$

$$80° = 100° - m\widehat{RT}$$

$$20° = m\widehat{RT}$$

10. 43.96 cm

$$\text{area of circle } = \pi r^2 \approx 3.14r^2$$

$$\frac{153.86}{3.14} = \frac{3.14r^2}{3.14}$$

$$49 = r^2$$

$$7 = r$$

$$C = 2\pi r = 2\pi(7) = 14\pi \approx 43.96 \text{ cm}$$

11. 2 times

To determine how many times the wheel will turn, find the circumference of the wheel.

$$C = \pi d$$

$$C = 20\pi$$

$$C \approx 62.8 \text{ inches}$$

The unicycle needs to travel 10 feet = 120 inches

$$\frac{120 \text{ inches}}{62.8 \text{ inches}} \approx 1.91$$

The wheel will turn approximately 2 times.

12. 3

The area of a semicircle is equal to half the area of its whole circle.

$$A = \frac{1}{2}\pi r^2$$

$$A = \frac{1}{2}\pi(15)^2$$

$$A = \frac{1}{2}(225\pi)$$

$$A \approx 353.25 \text{ square feet}$$

If each bag covers 150 square feet, Margaret needs to buy $\frac{353.25}{150} = 2.355$, which must be rounded up to 3 bags of seed.

13. 41.87 inches

$$\text{length of } \overparen{ABC} = \frac{300°}{360°}(2\pi(8))$$

$$\text{length of } \overparen{ABC} = \frac{5}{6}(16\pi)$$

$$\text{length of } \overparen{ABC} \approx 41.87 \text{ inches}$$

14. 57.7 cm²

$m\angle AOB + m\angle BOC = 180°$

$45° + m\angle BOC = 180°$

$m\angle BOC = 135°$

area of sector $BOC = \dfrac{m\angle BOC}{360°} \cdot \pi r^2$

area of sector $BOC = \dfrac{135°}{360°}(\pi(7)^2)$

area of sector $BOC = \dfrac{3}{8}(49\pi)$

area of sector $BOC \approx 57.7 \text{ cm}^2$

15. 10.5 inches

Since the circle is split into 12 equal arcs, the measure of each arc $= \dfrac{360°}{12} = 30°$.

Therefore, $m\widehat{AE} = 120°$ and $m\angle AOE = 120°$.

length of $\widehat{AE} = \dfrac{m\angle AOE}{360°} \cdot 2\pi r$

length of $\widehat{AE} = \dfrac{120°}{360°}(2\pi(5))$

length of $\widehat{AE} = \dfrac{1}{3}(10\pi)$

length of $\widehat{AE} \approx 10.5$ inches

16. 50.24 in^2

$$\text{area of fan} = \frac{160°}{360°}(\pi(6)^2)$$

$$\text{area of fan} \approx 50.24 \text{ in}^2$$

17. A

The formula is $(x - h)^2 + (y - k)^2 = r^2$

For this example, $h = 1$ and $k = -1$.

The center is at $(1, -1)$.

18. D

$(x - h)^2 + (y - k)^2 = r^2$

$(x - (-4))^2 + (y - 0)^2 = 4^2$

$(x + 4)^2 + y^2 = 16$

19. The center is at (–3, 2) and the radius is $\sqrt{9} = 3$.

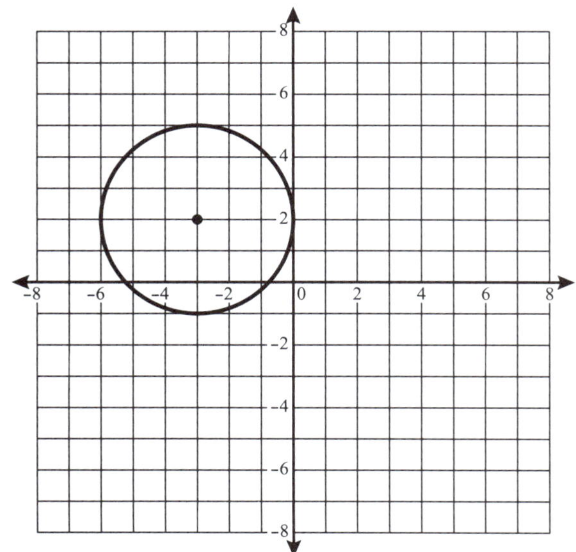

20. 201 square units

The circle $x^2 + y^2 = 64$ has a radius of 8 because $\sqrt{64} = 8$.

$A = \pi r^2$

$A = \pi(8)^2$

$A = 64\pi$

$A \approx 201$ square units

Florida Geometry EOC Practice Test

Also available at the REA Study Center (*www.rea.com/studycenter*)

This practice exam is also available at the REA Study Center. To closely simulate your test-day experience with the computer-based Florida EOC assessment, we suggest that you take the online version of the practice test. When you do, you'll also enjoy these benefits:

- Instant scoring
- Enforced time conditions
- Detailed score report of your strengths and weaknesses

Algebra 1 and Geometry End-of-Course Assessments Reference Sheet

Area

Parallelogram	$A = bh$
Triangle	$A = \frac{1}{2}bh$
Trapezoid	$A = \frac{1}{2}h(b_1 + b_2)$
Circle	$A = \pi r^2$
Regular Polygon	$A = \frac{1}{2}aP$

KEY

b = base	A = area
h = height	B = area of base
w = width	C = circumference
d = diameter	V = volume
r = radius	P = perimeter of base
ℓ = slant height	
a = apothem	$S.A.$ = surface area

Use 3.14 or $\frac{22}{7}$ for π.

Circumference

$C = \pi d$ or $C = 2\pi r$

		Volume/Capacity	Total Surface Area
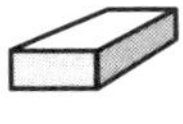	Rectangular Prism	$V = bwh$ or $V = Bh$	$S.A. = 2bh + 2bw + 2hw$ or $S.A. = Ph + 2B$
	Right Circular Cylinder	$V = \pi r^2 h$ or $V = Bh$	$S.A. = 2\pi rh + 2\pi r^2$ or $S.A. = 2\pi rh + 2B$
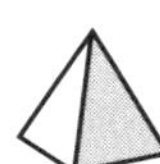	Right Square Pyramid	$V = \frac{1}{3}Bh$	$S.A. = \frac{1}{2}P\ell + B$
	Right Circular Cone	$V = \frac{1}{3}\pi r^2 h$ or $V = \frac{1}{3}Bh$	$S.A. = \frac{1}{2}(2\pi r)\ell + B$ or $S.A. = \pi r\ell + \pi r^2$
	Sphere	$V = \frac{4}{3}\pi r^3$	$S.A. = 4\pi r^2$

Sum of the measures of the interior angles of a polygon $= 180(n-2)$

Measure of an interior angle of a regular polygon $= \frac{180(n-2)}{n}$

where:

n represents the number of sides

Algebra 1 and Geometry End-of-Course Assessments Reference Sheet

Slope formula

$$m = \frac{y_2 - y_1}{x_2 - x_1}$$

where m = slope and (x_1, y_1) and (x_2, y_2) are points on the line

Slope-intercept form of a linear equation

$$y = mx + b$$

where m = slope and b = y-intercept

Point-slope form of a linear equation

$$y - y_1 = m(x - x_1)$$

where m = slope and (x_1, y_1) is a point on the line

Distance between two points

$P_1(x_1, y_1)$ and $P_2(x_2, y_2)$

$$\sqrt{(x_2 - x_1)^2 + (y_2 - y_1)^2}$$

Midpoint between two points

$P_1(x_1, y_1)$ and $P_2(x_2, y_2)$

$$\left(\frac{x_1 + x_2}{2}, \frac{y_1 + y_2}{2}\right)$$

Quadratic formula

$$x = \frac{-b \pm \sqrt{b^2 - 4ac}}{2a}$$

where a, b, and c are coefficients in an equation of the form $ax^2 + bx + c = 0$

Special Right Triangles

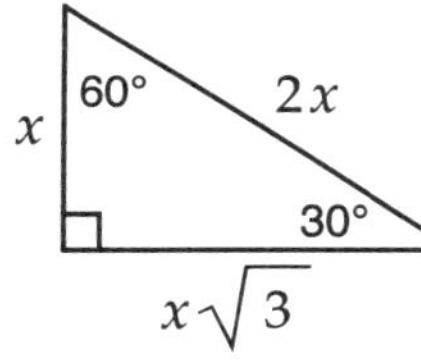

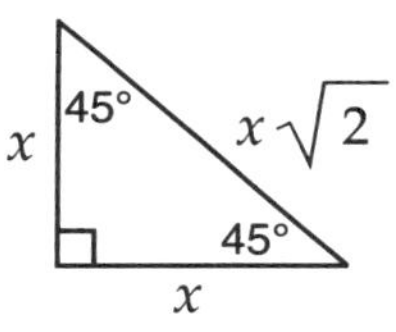

Trigonometric Ratios

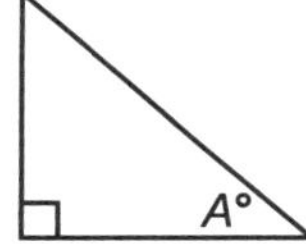

$\sin A° = \frac{\text{opposite}}{\text{hypotenuse}}$

$\cos A° = \frac{\text{adjacent}}{\text{hypotenuse}}$

$\tan A° = \frac{\text{opposite}}{\text{adjacent}}$

Conversions

1 yard = 3 feet
1 mile = 1,760 yards = 5,280 feet
1 acre = 43,560 square feet
1 hour = 60 minutes
1 minute = 60 seconds

1 cup = 8 fluid ounces
1 pint = 2 cups
1 quart = 2 pints
1 gallon = 4 quarts
1 pound = 16 ounces
1 ton = 2,000 pounds

1 meter = 100 centimeters = 1000 millimeters
1 kilometer = 1000 meters
1 liter = 1000 milliliters = 1000 cubic centimeters
1 gram = 1000 milligrams
1 kilogram = 1000 grams

Directions for Taking Geometry EOC Practice Test A

Test Questions

This Practice Test contains 51 questions. The number of questions on the actual test will vary.

- **Multiple-Choice Questions**

  Select the best answer for each question and mark it on the answer sheet on page 271.

- **Open-ended Questions**

  As you come to an open-ended question, use the Notes pages at the back of the book to do your work. Then fill in the answer using the digits 0–9 and/or the symbols for a decimal point, fraction bar, or negative sign in the answer box provided for each specific open-ended question.

Reference Pages

You may refer to the two preceding Reference Pages as often as you like.

Timing

For the actual test you will be given two 80-minute periods to complete the test, with a ten-minute break in between. However, anyone who has not finished will be allowed to continue working.

Checking Your Answers

You will find the correct answers, along with detailed explanations, for this practice test beginning on page 273.

Reviewing Your Work

When finished, turn to the grid on page 326. Circle the number of any questions that you missed in Test A. You will be able to see a pattern that shows which Benchmarks will need your further attention.

1 Lily traveled 10 blocks north and 5 blocks east to visit her friend Marie. If she were able to walk directly to Marie's house, approximately how many blocks would Lily travel?

A. 5

B. 8.7

C. 11.2

D. 15

2 Determine the value of y.

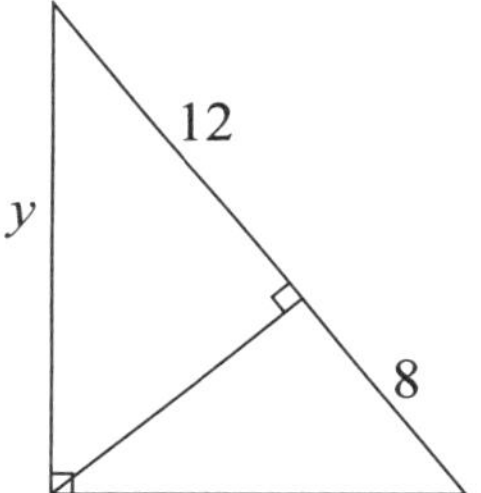

A. 8

B. 9.8

C. 15.5

D. 20

Go On

3 The scale factor of $\triangle ABC$ to $\triangle XYZ$ is $\frac{1}{5}$. If the area of $\triangle ABC$ is 8 square centimeters (cm^2), what is the area of $\triangle XYZ$, in cm^2?

4 $\triangle ABC \sim \triangle DEF$. Which of the following is the scale factor?

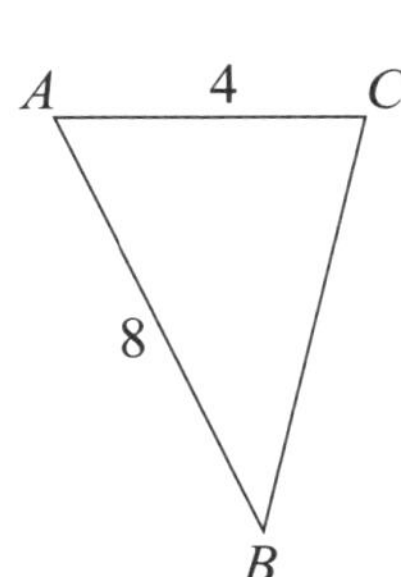

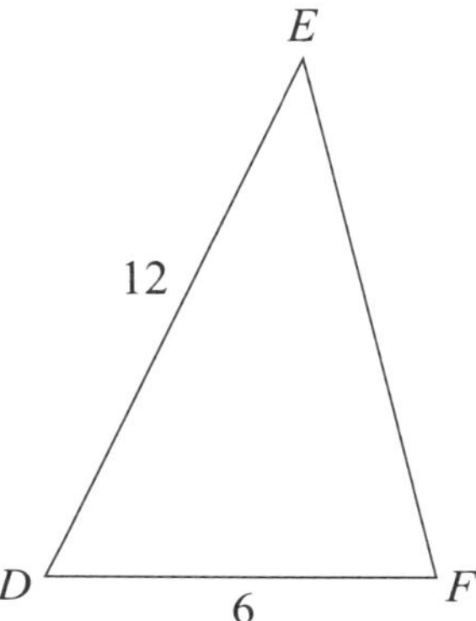

A. $\frac{1}{3}$

B. $\frac{1}{2}$

C. $\frac{2}{3}$

D. $\frac{3}{4}$

Go On

5 Given $\triangle ABC \cong \triangle DEF$, which of the following statements is NOT true?

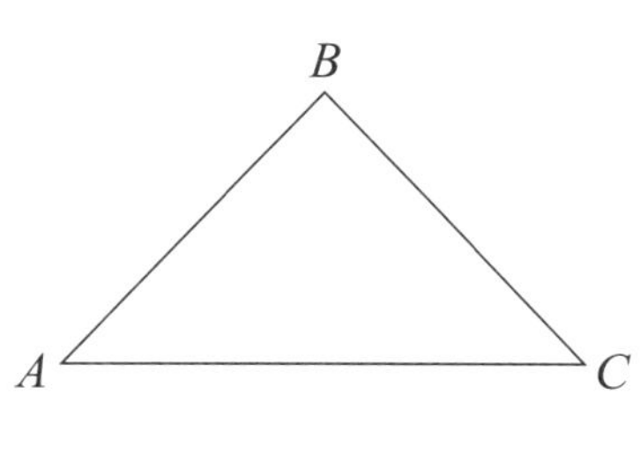

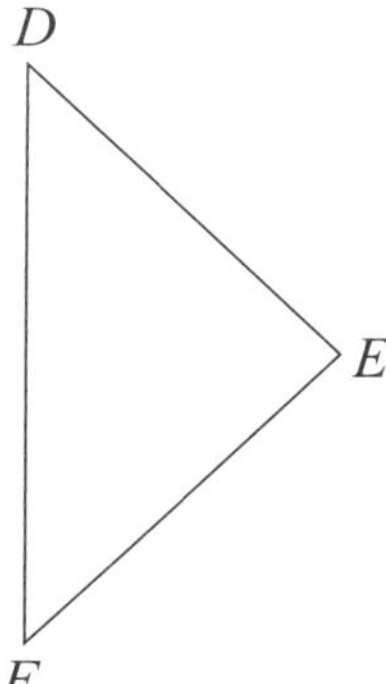

A. $m\angle A = m\angle D$

B. $m\angle B = m\angle E$

C. $AB = DE$

D. $BC = DF$

6 Which of the following triangles is impossible to construct?

A. equilateral right

B. scalene right

C. isosceles right

D. isosceles obtuse

7 Find the measure, in degrees, of an interior angle of a regular pentagon.

Go On

8 Jessie is 5 feet tall and she casts a shadow of 12 feet. If a tree casts a 30-foot shadow, how tall is the tree in feet? Do NOT round off your answer.

9 Given that *ABCDEF* is similar to *UVWXYZ*, what is the length of $\overline{XY}$?

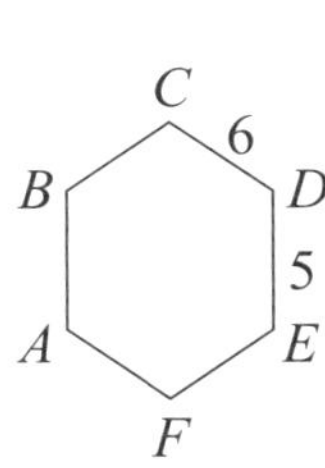

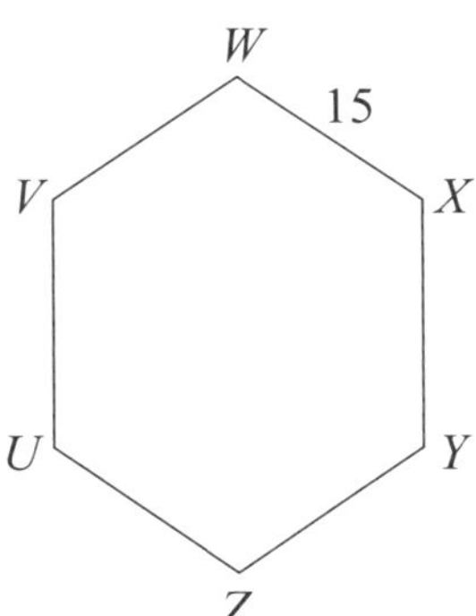

A. 5

B. 12.5

C. 16

D. 18

Go On

10 Which of the following is the distance between (2, –4) and (–4, 6)?

A. $2\sqrt{10}$

B. $2\sqrt{26}$

C. $2\sqrt{34}$

D. 4

11 M (5, –3) is the midpoint of $\overline{AB}$, where A (7, y) and B (3, –4). What is the y-coordinate of point A?

12 Find $m\angle 1$.

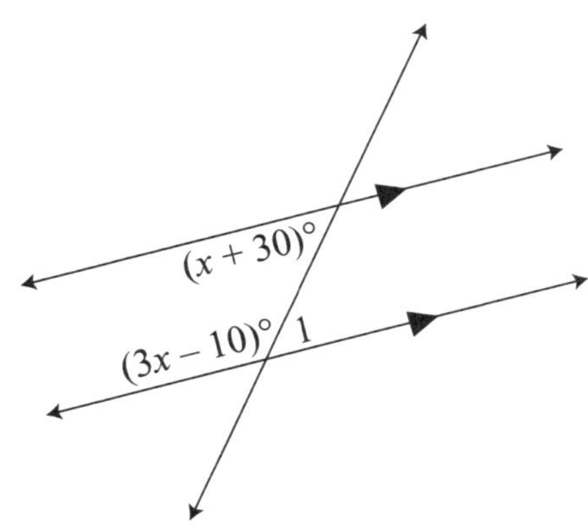

A. 40°

B. 70°

C. 110°

D. 140°

Go On

13 In the figure below, $\overleftrightarrow{XY}$ is a ______________.

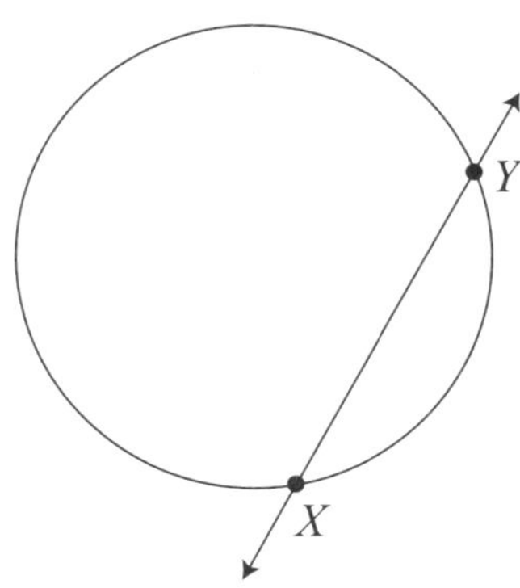

A. radius

B. chord

C. secant

D. tangent

14 Brynn rode her bicycle down her street. If the front wheel rotated 25 times, approximately how far did she travel?

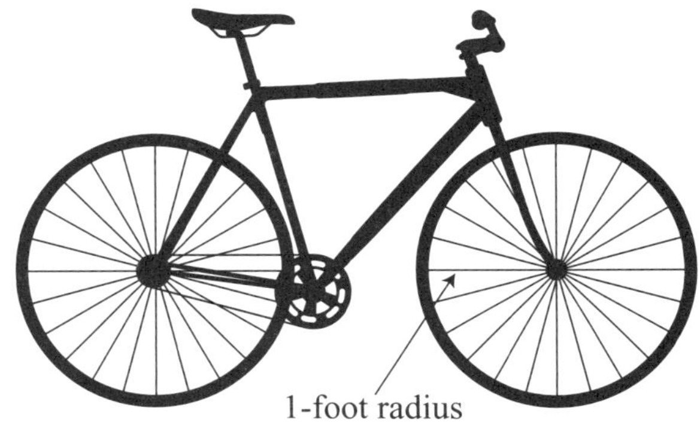

A. 78.5 feet

B. 157 feet

C. 235.5 feet

D. 314 feet

15 A large pizza has a diameter of 18 inches. The pizza is cut into 8 equal pieces and Mandy ate 2 slices. To the nearest tenth, what is the area in square inches of the pizza she ate?

16 Find $m\widehat{AC}$.

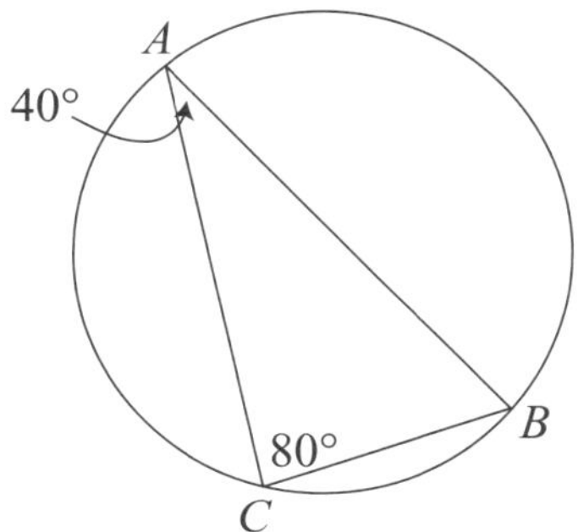

A. 60°

B. 80°

C. 120°

D. 160°

17 Find $m\angle ACE$ in degrees.

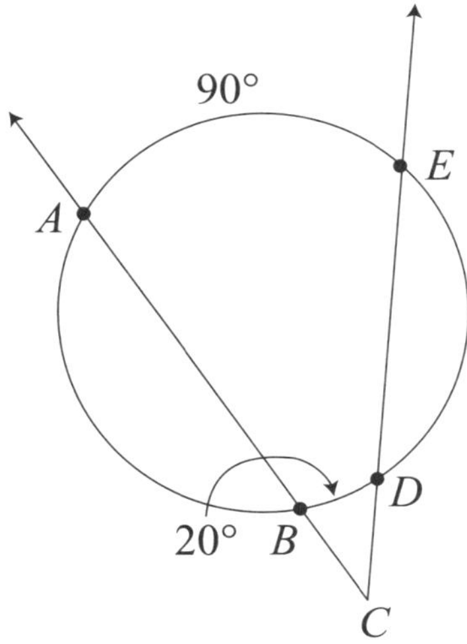

18 Which of the following is the center of the circle represented by the equation $(x - 4)^2 + (y + 9)^2 = 121$?

A. (4, 9)

B. (4, –9)

C. (–4, 9)

D. (–4, –9)

19 Which of the following is the graph of $(x - 1)^2 + (y - 3)^2 = 9$?

A.

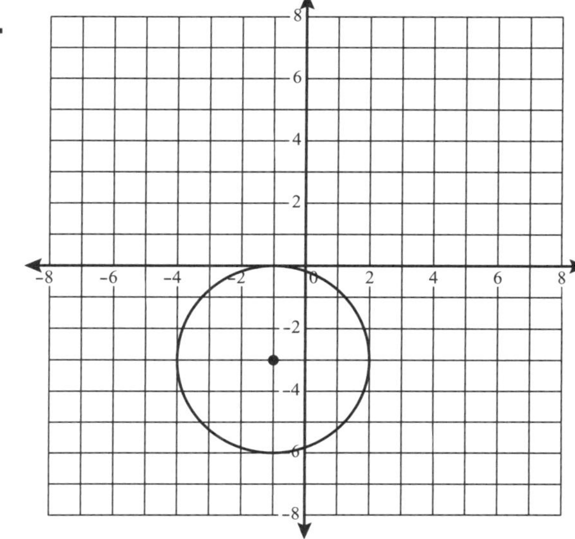

B.

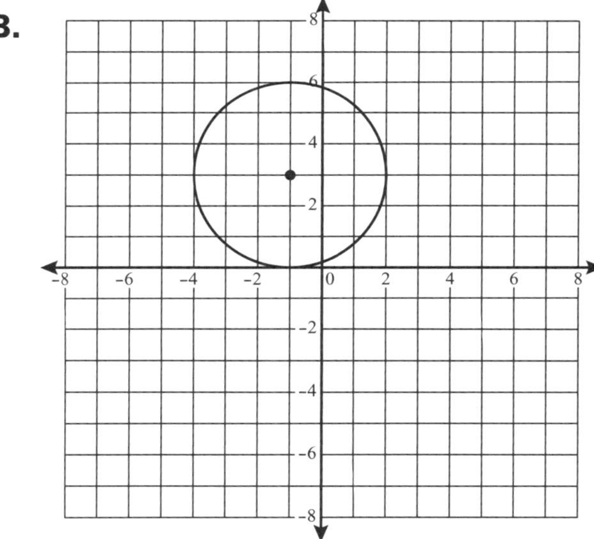

C.

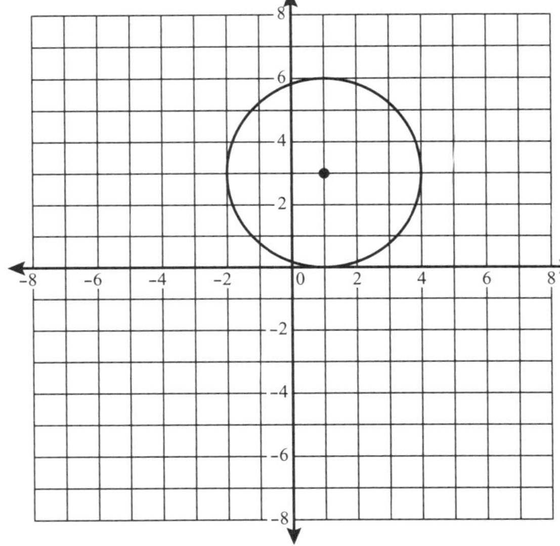

D.

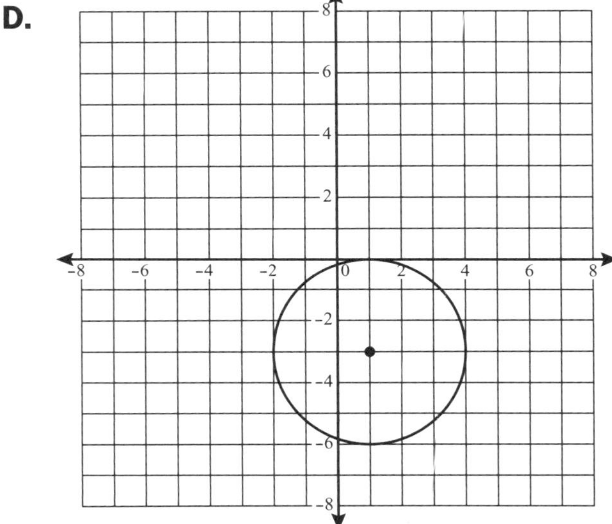

Go On

20 Television screens are measured by their diagonal. Below is a 40-inch (in.) television. Find the length of the television screen in inches.

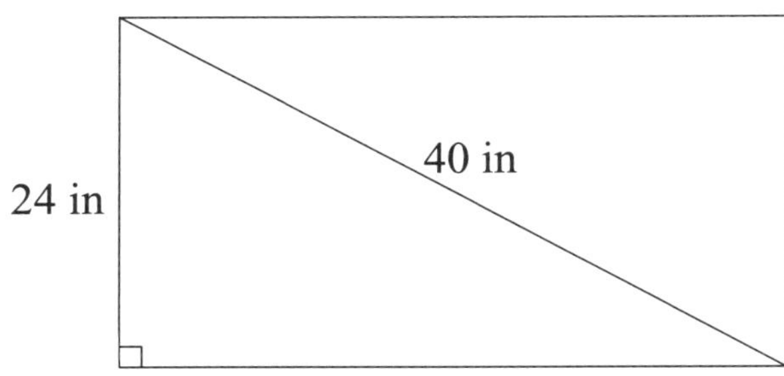

21 Which of the following is NOT the net of a cube?

A. 

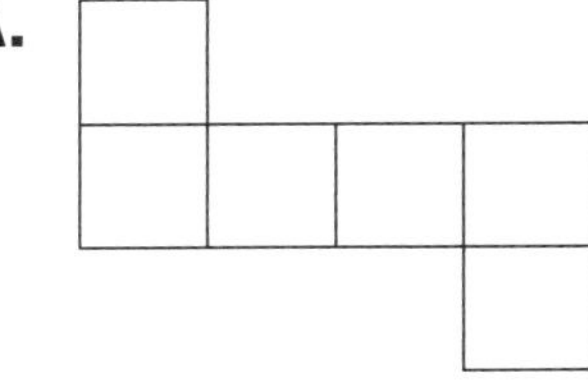

B.

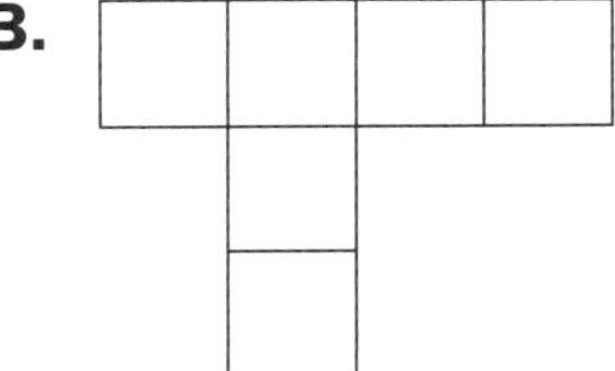

C.

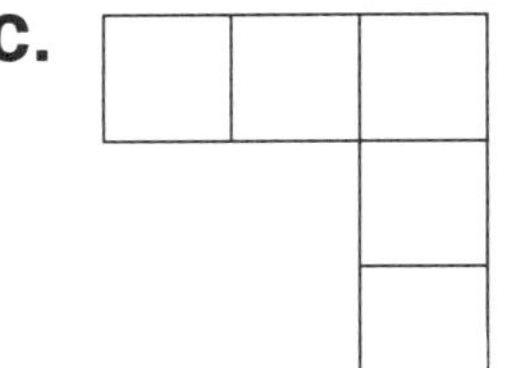

D.

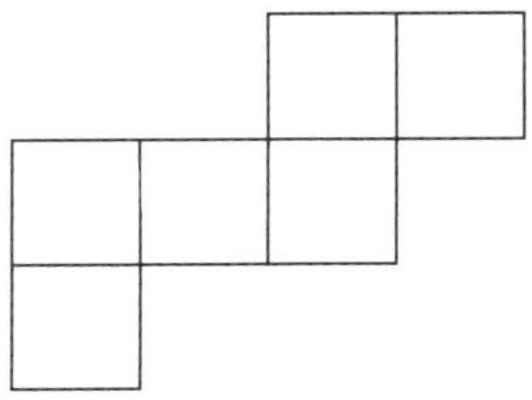

Go On

22 Determine the number of vertices in a solid figure that has 8 faces and 14 edges.

A. 2

B. 8

C. 18

D. 24

23 Which of the following shows the image of $\triangle ABC$ after a reflection over the x-axis?

A.

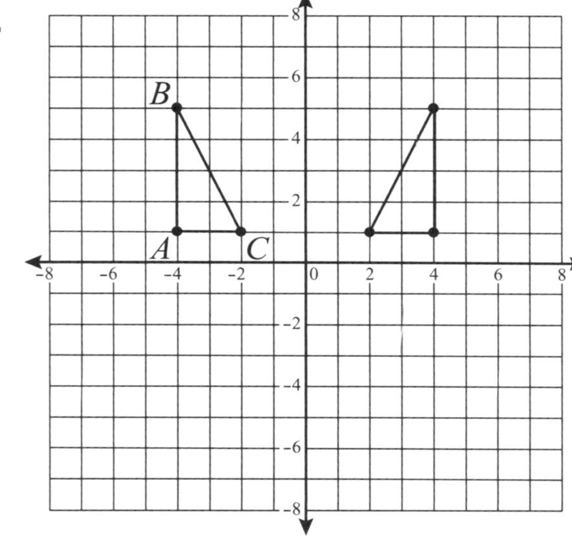

Go On

B.

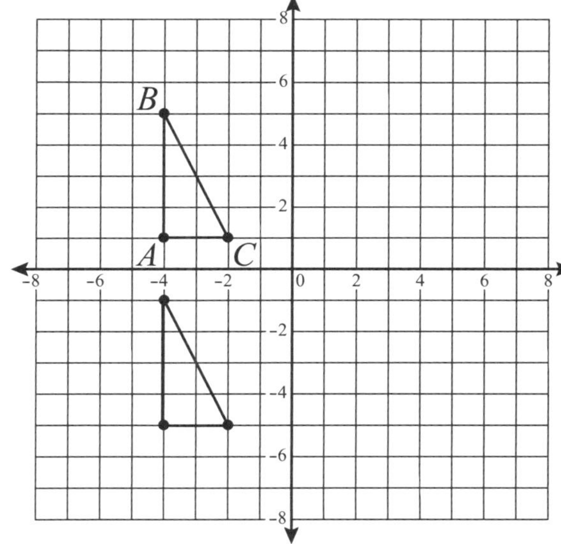

C.

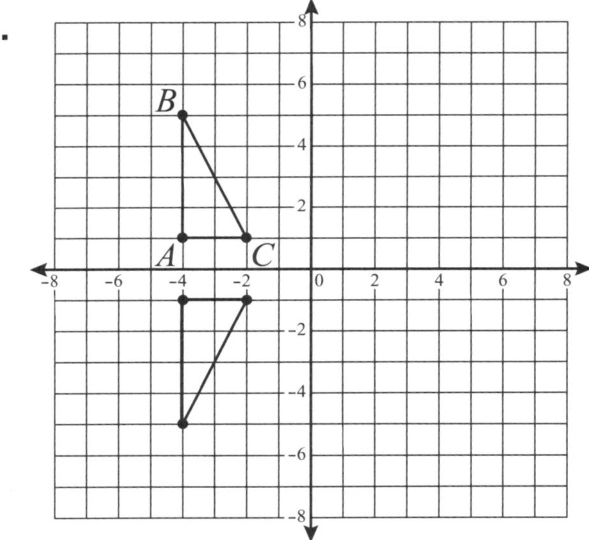

D.

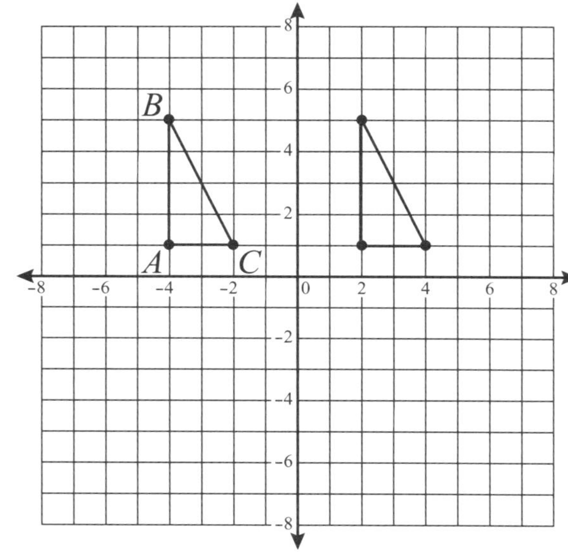

24 Which of the following describes the translation that results in $\triangle D'E'F'$?

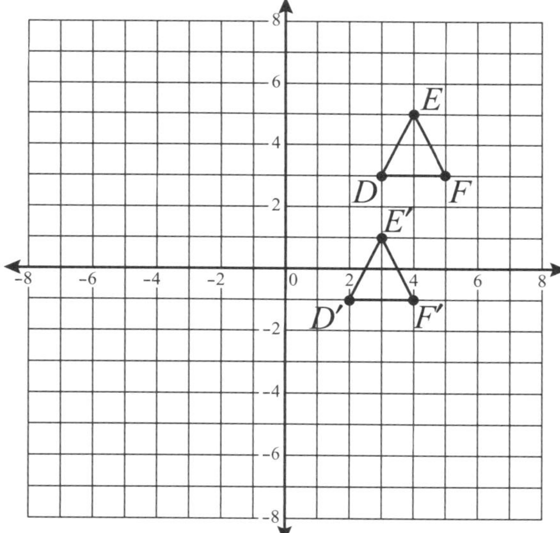

A. $(x, y) \rightarrow (x - 1, y + 4)$

B. $(x, y) \rightarrow (x - 1, y - 4)$

C. $(x, y) \rightarrow (x + 4, y - 1)$

D. $(x, y) \rightarrow (x - 4, y - 1)$

25 A 4-foot square is cut out of a 9-foot square piece of wood. What is the area of the remaining wood in square feet?

26 Which expression represents the perimeter of the triangle?

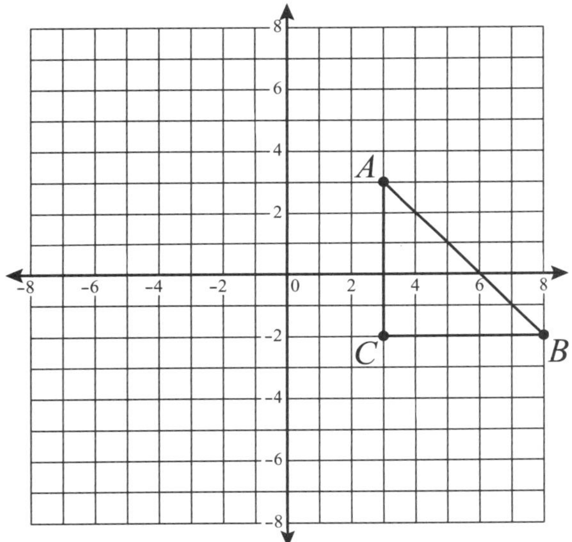

A. $10+\sqrt{2}$

B. $10+5\sqrt{2}$

C. $15+\sqrt{2}$

D. $15\sqrt{2}$

27 Which of the following is NOT an undefined term of geometry?

A. point

B. segment

C. line

D. plane

28 Carlos needs to calculate the surface area of a sphere with diameter of 10 feet. He guesses that the answer is 300 ft^2. Using $\pi = 3.14$, determine the positive difference in square feet between the actual surface area and Carlos's guess.

29 In $\triangle ABC$, angle C is a right angle. If $\sin A = \frac{3}{5}$, what is $\tan B$? Write your answer in fraction form.

Go On

30 Figure $ABCD$ is a quadrilateral. By definition, what additional information is necessary to state that it is a parallelogram?

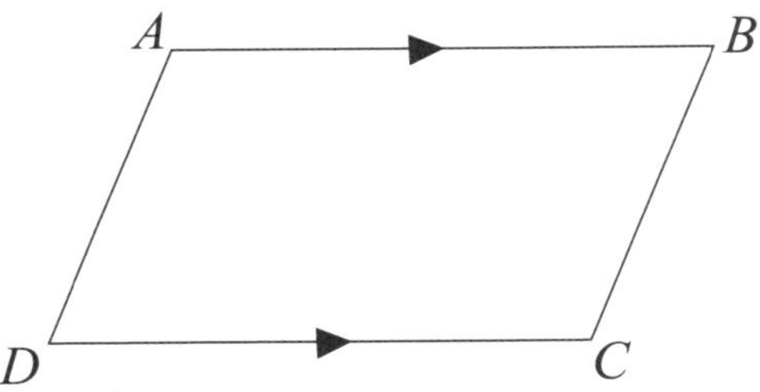

A. $AB = CD$

B. $AD = BC$

C. $\overline{AD} \parallel \overline{BC}$

D. $m\angle A = m\angle C$

31 What is the contrapositive of the following statement?

If two lines are parallel, then the lines never intersect.

A. If two lines never intersect, then the lines are parallel.

B. If two lines are not parallel, then the lines will intersect.

C. If two lines intersect, then the lines are not parallel.

D. If two lines are not parallel, then the lines never intersect.

Go On

32 Which statement best explains why the equation $2x = 4x - 10$ can be used to solve for x?

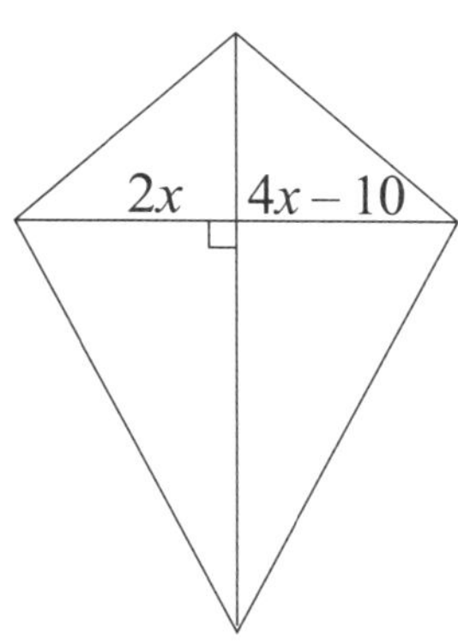

A. Diagonals of a kite are perpendicular

B. Diagonals of a kite bisect each other

C. The longer diagonal bisects the shorter diagonal

D. A kite is a type of quadrilateral

33 Which of the following are logically equivalent?

A. $p \vee q$ and $p \vee {\sim}q$

B. ${\sim}(p \vee q)$ and ${\sim}p \wedge {\sim}q$

C. ${\sim}(p \vee q)$ and ${\sim}p \vee {\sim}q$

D. ${\sim}(p \wedge q)$ and ${\sim}p \wedge {\sim}q$

34 Which of the following quadrilaterals has only one pair of parallel sides?

A. kite

B. trapezoid

C. rhombus

D. rectangle

35 If the rectangle on the grid below is translated seven units to the left and six units down, what will be the new x-coordinate of point B?

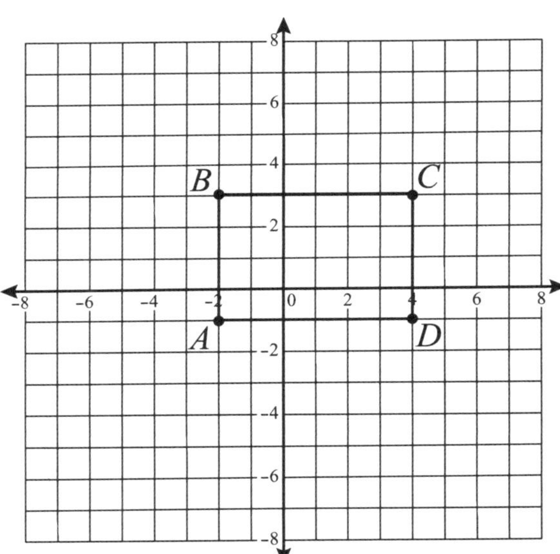

36 The surface area of a sphere is 144π square feet. What is the area of the sphere's great circle?

A. 6π square feet

B. 12π square feet

C. 36π square feet

D. 144π square feet

37 Perry has a 3-inch by 5-inch picture of his family. He wants to enlarge the picture to 5 times its original size. What is the perimeter of the enlarged picture in inches?

38 Find the surface area to the nearest square centimeter.

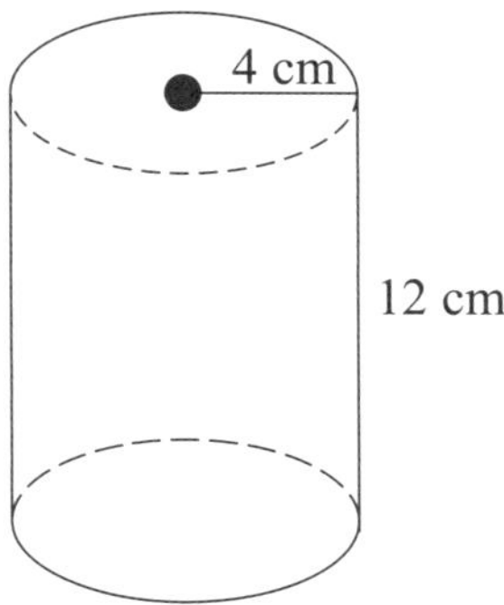

39 Find the volume of the square pyramid in cubic centimeters (cm^3).

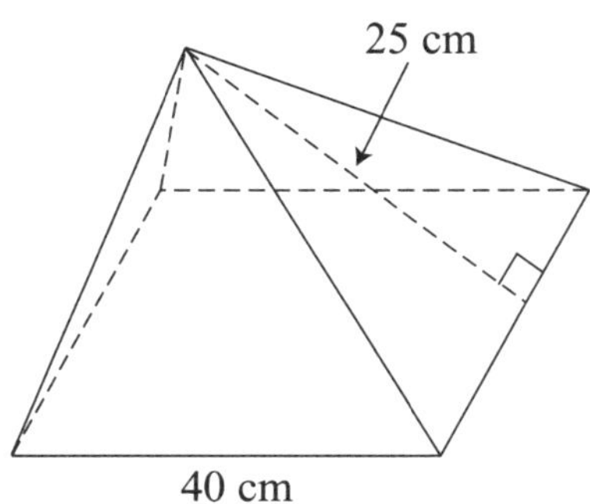

40 Given the figures are similar, what is the height of the larger cylinder?

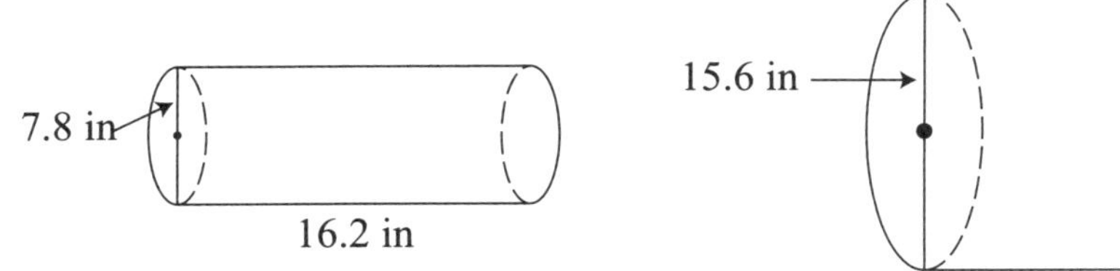

A. 24 inches

B. 31.2 inches

C. 32.4 inches

D. 46.8 inches

41 Two similar pyramids have a scale factor of $\frac{2}{5}$. If the volume of the smaller pyramid is 24cubic feet, what is the volume of the larger pyramid in cubic feet?

42 Determine the value of x.

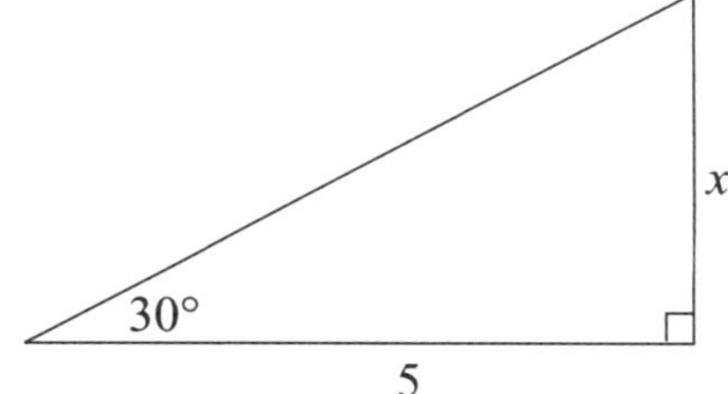

A. 5

B. $\frac{5\sqrt{3}}{3}$

C. $5\sqrt{3}$

D. 10

43 Given an isosceles right triangle with two legs measuring 7 feet, what is the length of the hypotenuse to the nearest tenth?

44 Figure A is formed by the points (0, –2), (–2, 0), (–5, 0), and (–7, –2) and three of the points of Figure B are (0, 3), (0, 6), and (2, 1). Given that these two figures are congruent, what are the coordinates of the missing point of Figure B?

A. (2, 8)

B. (8, 2)

C. (–2, 8)

D. (–8, 2)

45 A ladder is leaning against the side of a building. If the base of the ladder is 9 feet from the building and it forms a 50° angle with the ground, what is the approximate length of the ladder in feet?

A. 23 ft

B. 20 ft

C. 17 ft

D. 14 ft

46 M is the midpoint of NL and P is the centroid of $\triangle NOL$. If $MO = 15$, find the length of MP.

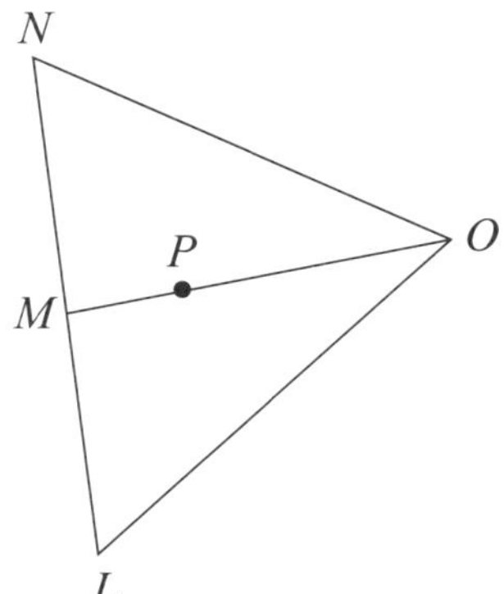

A. 5

B. 7

C. 8

D. 10

Go On

47 Which type of proof begins by assuming the statement to be proven is false?

A. Direct proof

B. Indirect proof

C. Paragraph proof

D. Two-column proof

48 Which of the following set of lengths cannot form a triangle?

A. 3, 4, 6

B. 8, 10, 12

C. 4, 4, 7

D. 2, 5, 8

49 Find the length of XY.

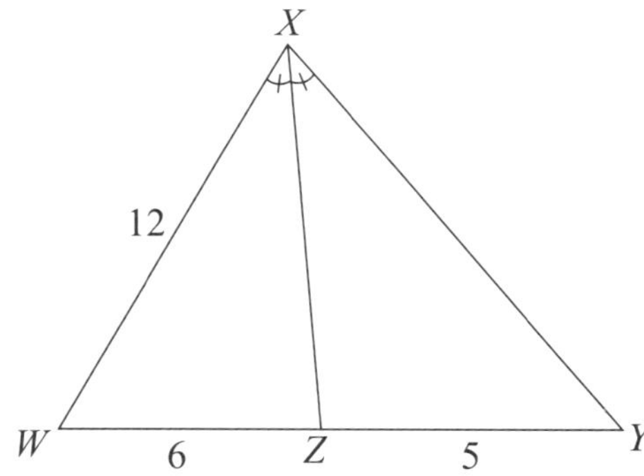

50 In the triangle below, which side is the shortest?

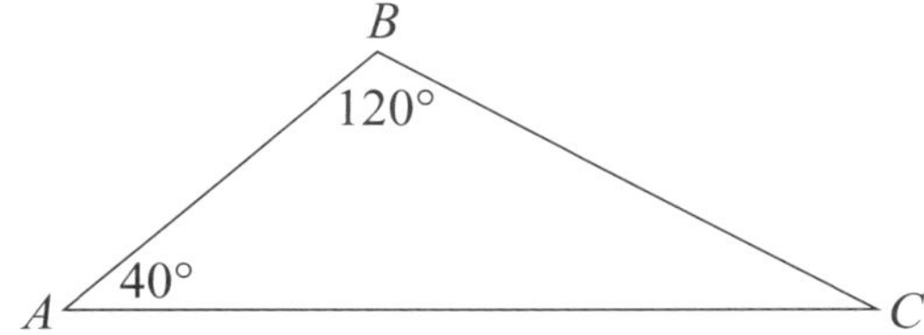

A. *AB*

B. *BC*

C. *AC*

D. Cannot be determined

51 Which of the following is a regular polygon?

A. parallelogram

B. rectangle

C. isosceles triangle

D. square

Go On

This is the end of the Geometry Test.

Until time is called, go back and check your work or answer questions you did not complete. When you have finished, close your Test Book and Answer Book.

Geometry Practice Test Form A

Answer Sheet

1.	Ⓐ Ⓑ Ⓒ Ⓓ	18.	Ⓕ Ⓖ Ⓗ Ⓘ	35.	☐☐☐☐☐☐☐
2.	Ⓐ Ⓑ Ⓒ Ⓓ	19.	Ⓐ Ⓑ Ⓒ Ⓓ	36.	Ⓐ Ⓑ Ⓒ Ⓓ
3.	☐☐☐☐☐☐☐	20.	☐☐☐☐☐☐☐	37.	☐☐☐☐☐☐☐
4.	Ⓐ Ⓑ Ⓒ Ⓓ	21.	Ⓐ Ⓑ Ⓒ Ⓓ	38.	☐☐☐☐☐☐☐
5.	Ⓐ Ⓑ Ⓒ Ⓓ	22.	Ⓐ Ⓑ Ⓒ Ⓓ	39.	☐☐☐☐☐☐☐
6.	Ⓐ Ⓑ Ⓒ Ⓓ	23.	Ⓐ Ⓑ Ⓒ Ⓓ	40.	Ⓐ Ⓑ Ⓒ Ⓓ
7.	☐☐☐☐☐☐☐	24.	Ⓐ Ⓑ Ⓒ Ⓓ	41.	☐☐☐☐☐☐☐
8.	☐☐☐☐☐☐☐	25.	☐☐☐☐☐☐☐	42.	Ⓐ Ⓑ Ⓒ Ⓓ
9.	Ⓐ Ⓑ Ⓒ Ⓓ	26.	Ⓐ Ⓑ Ⓒ Ⓓ	43.	☐☐☐☐☐☐☐
10.	Ⓐ Ⓑ Ⓒ Ⓓ	27.	Ⓐ Ⓑ Ⓒ Ⓓ	44.	Ⓐ Ⓑ Ⓒ Ⓓ
11.	☐☐☐☐☐☐☐	28.	☐☐☐☐☐☐☐	45.	Ⓐ Ⓑ Ⓒ Ⓓ
12.	Ⓐ Ⓑ Ⓒ Ⓓ	29.	☐☐☐☐☐☐☐	46.	Ⓐ Ⓑ Ⓒ Ⓓ
13.	Ⓐ Ⓑ Ⓒ Ⓓ	30.	Ⓐ Ⓑ Ⓒ Ⓓ	47.	Ⓐ Ⓑ Ⓒ Ⓓ
14.	Ⓐ Ⓑ Ⓒ Ⓓ	31.	Ⓐ Ⓑ Ⓒ Ⓓ	48.	Ⓐ Ⓑ Ⓒ Ⓓ
15.	☐☐☐☐☐☐☐	32.	Ⓐ Ⓑ Ⓒ Ⓓ	49.	☐☐☐☐☐☐☐
16.	Ⓐ Ⓑ Ⓒ Ⓓ	33.	Ⓐ Ⓑ Ⓒ Ⓓ	50.	Ⓐ Ⓑ Ⓒ Ⓓ
17.	☐☐☐☐☐☐☐	34.	Ⓐ Ⓑ Ⓒ Ⓓ	51.	Ⓐ Ⓑ Ⓒ Ⓓ

Answers
Practice Test

1 **C**

$a^2 + b^2 = c^2$

$10^2 + 5^2 = c^2$

$100 + 25 = c^2$

$\sqrt{125} = \sqrt{c^2}$

$11.2 \approx c$

2 **C**

$\frac{12}{y} = \frac{y}{20}$

$y^2 = 240$

$y \approx 15.5$

3 **200**

$\frac{1^2}{5^2} = \frac{8}{x}$

$\frac{1}{25} = \frac{8}{x}$

$x = 200$

4 **C**

$\frac{AC}{DF} = \frac{4}{6} = \frac{2}{3}$

$\frac{AB}{DE} = \frac{8}{12} = \frac{2}{3}$

5 **D**

$\overline{BC}$ corresponds to $\overline{EF}$. Therefore, $BC \neq DF$. Instead, $BC = EF$.

6 **A**

An equilateral triangle cannot have a right angle. By definition, all angles must be congruent. Therefore, each angle equals $180 \div 3 = 60°$.

7 **108**

A regular pentagon has five congruent angles. Using the formula $(n - 2)180°$, substitute 5 sides for n.

$(5 - 2)180° = (3)(180°) = 540°$. Then each interior angle $= \frac{540°}{5} = 108°$.

8 **12.5**

$$\frac{\text{Jessie's height}}{\text{shadow}} = \frac{\text{tree height}}{\text{shadow}}$$

$$\frac{5}{12} = \frac{x}{30}$$

$$150 = 12x$$

$$12.5 = x$$

9 **B**

Let x represent XY.

$$\frac{CD}{WX} = \frac{DE}{XY}$$

$$\frac{6}{15} = \frac{5}{x}$$

$$6x = 75$$

$$x = 12.5$$

10 **C**

$$d = \sqrt{(x_2 - x_1)^2 + (y_2 - y_1)^2}$$

$$d = \sqrt{(-4 - 2)^2 + (6 - (-4))^2}$$

$$d = \sqrt{(-6)^2 + (10)^2}$$

$$d = \sqrt{36 + 100}$$

$$d = \sqrt{136} = 2\sqrt{34}$$

11 **–2**

$$M = \left(\frac{x_2 + x_1}{2}, \frac{y_2 + y_1}{2} \right)$$

$$(5, -3) = \left(\frac{7 + 3}{2}, \frac{-4 + y_1}{2} \right)$$

Since the value of y_1 is unknown, the following equation can be used to solve for y_1:

$$-3 = \frac{-4 + y_1}{2}$$

$$-6 = -4 + y_1$$

$$-2 = y_1$$

12 **B**

Since $\angle 1$ and the angle that measures $(x + 30)°$ are alternate interior angles, they are congruent. Therefore $\angle 1 = (x + 30)°$. The angle that measures $(3x - 10)°$ and $\angle 1$ are a linear pair. Therefore, $(x + 30)° + (3x - 10)° = 180° \Rightarrow (4x + 20)° = 180° \Rightarrow 4x = 160° \Rightarrow x = 40°$. Since $m\angle 1 = (x + 30)°$, substitute $x = 40 \Rightarrow m\angle 1 = 40° + 30° = 70°$.

13 **C**

$\overleftrightarrow{XY}$ is a line that intersects the circle twice. Therefore, it is a secant.

14 **B**

If the wheel rotates 25 times, the bicycle traveled 25 times the circumference of the wheel. $C = 2\pi r = 2\pi(1) \approx 6.28 \Rightarrow (25)(6.28) \approx 157$ feet.

15 **63.6**

If the pizza is cut into 8 equal pieces, each piece is $\frac{360°}{8} = 45°$ of the pizza. Therefore, 2 slices is 90° of the pizza. The slices represent a sector of the pizza. So, the area of the two slices equals $\frac{90°}{360°}(\pi)(9^2) \approx 63.6 \text{ in}^2$.

16 **C**

$m\overset{\frown}{AC} = 2(m\angle ABC)$. $m\angle ABC = 180° - 40° - 80° = 60°$. Therefore, $m\overset{\frown}{AC} = 120°$.

17 **35**

$m\angle ACE = \frac{1}{2}(m\overset{\frown}{AE} - m\overset{\frown}{BD}) = \frac{1}{2}(90° - 20°) = \frac{1}{2}(70°) = 35°$.

18 **B**

The standard equation of a circle is $(x - h)^2 + (y - k)^2 = r^2$, where (h, k) is the center. Therefore, the center is (4, –9).

19 **C**

The center of the circle $(x - 1)^2 + (y - 3)^2 = 9$ is located at (1, 3) with a radius of 3. This is represented by graph C.

20 **32**

$a^2 + b^2 = c^2$

$24^2 + b^2 = 40^2 \Rightarrow 576 + b^2 = 1600 \Rightarrow \sqrt{b^2} = \sqrt{1024} \Rightarrow b = 32$.

21 **C**

A cube has six faces. The drawing in C represents only five faces, so it cannot be the net of a cube.

22 **B**

Euler's formula states: $V - E + F = 2 \Rightarrow V - 14 + 8 = 2 \Rightarrow V - 6 = 2 \Rightarrow V = 8$.

23 **C**

When a figure is reflected over the x-axis, a point (x, y) becomes $(x, -y)$. For example, point A is located at $(-4, 1)$ and the corresponding point A' is located at $(-4, -1)$.

24 **B**

$\triangle DEF$ is translated 1 unit left and 4 units down.

25 **65**

The original wood has an area of 81 square feet and a 16 square foot hole is cut out of it. The remaining area is found by subtracting $\Rightarrow 81 - 16 = 65$ square feet.

26 **B**

$AC = 5$ and $BC = 5$. Using the rules of 45°-45°-90° triangles, $AB = 5\sqrt{2}$. The perimeter is $5 + 5 + 5\sqrt{2} = 10 + 5\sqrt{2}$.

27 **B**

A segment is defined as a portion of a line with two endpoints.

28 **14**

Surface area of a sphere is $4\pi r^2$. The radius is 5 feet. So the actual surface area, using $\pi = 3.14$, is $(4)(3.14)(25) = 314$. The difference between this answer and Carlos's guess is 14 ft^2.

29 $\frac{4}{3}$

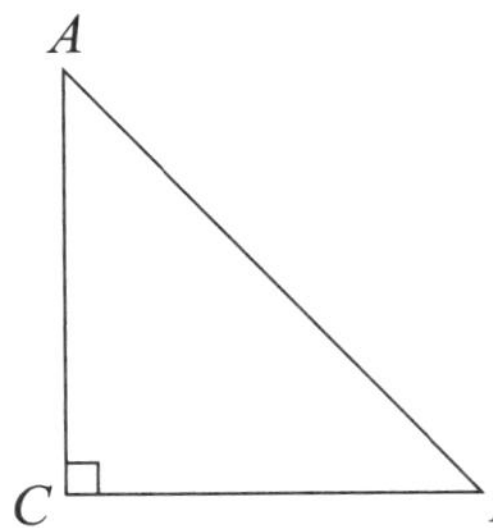

If $\sin A = \frac{3}{5}$ and $\sin A = \frac{\text{opposite}}{\text{hypotenuse}}$, then $BC = 3$ and $AB = 5$. Using the Pythagorean Theorem, $AC = 4$. So $\tan B = \frac{\text{opposite}}{\text{adjacent}} = \frac{4}{3}$.

30 **C**

The definition of a parallelogram is a quadrilateral with two pairs of parallel sides.

31 **C**

The contrapositive switches the hypothesis and conclusion and negates both parts of the statement.

32 **C**

The diagonals of a kite are perpendicular to each other and the longer diagonal bisects the shorter diagonal.

33 **B**

p	q	$\sim p$	$\sim q$	$p \vee q$	$p \wedge q$	$\sim p \vee \sim q$	$\sim p \wedge \sim q$	$\sim(p \vee q)$
T	T	F	F	T	T	F	F	F
T	F	F	T	T	F	T	F	F
F	T	T	F	T	F	T	F	F
F	F	T	T	F	F	T	T	T

34 **B**

By definition, a trapezoid has only one pair of parallel sides.

35 **–9**

The current x-coordinate of point B is –2. After the rectangle is translated seven units to the left, its x-coordinate becomes $-2 - 7 = -9$. The rectangle will then be translated six units down, but the x-coordinate will not be affected.

36 **C**

Surface Area $= 4\pi r^2$; therefore, $144\pi = 4\pi r^2 \Rightarrow 36 = r^2 \Rightarrow r = 6$ feet. The area of the great circle is $\pi r^2 \Rightarrow \pi(6)^2 = 36\pi$ square feet.

37 **80**

The perimeter of the original photo is $3 + 3 + 5 + 5 = 16$ inches. If the picture is enlarged to 5 times the size, the perimeter is also 5 times as large $\Rightarrow$ 5(16 inches) = 80 inches.

38 **402**

Surface area of a cylinder $= 2\pi r^2 + 2\pi rh \Rightarrow 2\pi(4)^2 + 2\pi(4)(12) = 32\pi + 96\pi = 128\pi$ cm$^2 \approx (128)(3.14) \approx 402$ cm^2.

39 **8000**

Volume of a pyramid $\frac{1}{3}Bh$. The area of the square base is $(40)^2 = 1600$ cm^2. To find the height, use the Pythagorean Theorem. $a^2 + b^2 = c^2 \Rightarrow 20^2 + b^2 = 25^2 \Rightarrow 400 + b^2 = 625\ \ b^2 = 225$ so $b = 15$. So, the height is 15 cm. $V = \frac{1}{3}(1600)(15) = 8000$ cm^3.

40 **C**

$$\frac{7.8}{15.6} = \frac{16.2}{x}$$

$$7.8x = 252.72$$

$$x = 32.4$$

41 **375**

$$\frac{2^3}{5^3} = \frac{24}{x}$$

$$\frac{8}{125} = \frac{24}{x}$$

$$8x = 3000$$

$$x = 375$$

42 **B**

Using the rules of 30°-60°-90° triangles, $x = \frac{5}{\sqrt{3}}$. Rationalize the denominator and $x = \frac{5}{\sqrt{3}} \cdot \frac{\sqrt{3}}{\sqrt{3}} = \frac{5\sqrt{3}}{3}$.

43 **9.9**

Using the rules of 45°-45°-90° triangles, the legs are represented by a and the hypotenuse is $a\sqrt{2}$. Since the legs measure 7 feet, the hypotenuse is $7\sqrt{2}$ feet ≈ 9.9 feet.

44 **A**

Using the point (2, 8), each trapezoid will have bases of 3 units and 7 units, and a height of 2 units.

45 **D**

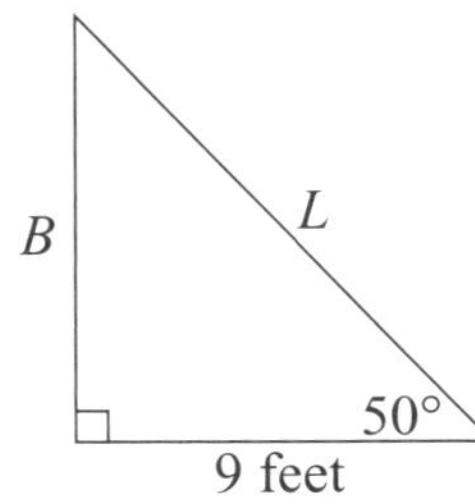

$$\tan 50° = \frac{B}{9}$$

$$1.19 = \frac{B}{9}$$

$$10.71 \approx B$$

$$a^2 + b^2 = c^2$$

$$9^2 + 10.71^2 = c^2$$

$$81 + 114.7 = c^2$$

$$195.7 = c^2$$

$$14 \approx c$$

46 **A**

$PO = \frac{2}{3} MO$

$PO = \frac{2}{3}(15)$

$PO = 10$

$MP + PO = MO$

$MP + 10 = 15$

$MP = 5$

47 **B**

An indirect proof assumes the prove statement is false.

48 **D**

By the Triangle Inequality theorem, the sum of the lengths of any two sides must exceed the length of the third side.

$2 + 5 \ngtr 8$

49 **10**

$\frac{WX}{WZ} = \frac{YX}{YZ}$

$\frac{12}{6} = \frac{x}{5}$

$60 = 6x \Rightarrow x = 10$

50 **A**

$m\angle C = 20°$. Since $\angle C$ is the smallest angle, the side opposite $\angle C$ is the shortest. $\overline{AB}$ is opposite $\angle C$.

51 **D**

All sides of a square are congruent and all angles are congruent. Therefore, it is a regular polygon.

Florida Geometry EOC Practice Test

Also available at the REA Study Center (*www.rea.com/studycenter*)

This practice exam is also available at the REA Study Center. To closely simulate your test-day experience with the computer-based Florida EOC assessment, we suggest that you take the online version of the practice test. When you do, you'll also enjoy these benefits:

- Instant scoring
- Enforced time conditions
- Detailed score report of your strengths and weaknesses

Algebra 1 and Geometry End-of-Course Assessments Reference Sheet

Area

Parallelogram	$A = bh$
Triangle	$A = \frac{1}{2}bh$
Trapezoid	$A = \frac{1}{2}h(b_1 + b_2)$
Circle	$A = \pi r^2$
Regular Polygon	$A = \frac{1}{2}aP$

KEY

b = base	A = area
h = height	B = area of base
w = width	C = circumference
d = diameter	V = volume
r = radius	P = perimeter of base
ℓ = slant height	
a = apothem	$S.A.$ = surface area

Use 3.14 or $\frac{22}{7}$ for π.

Circumference

$C = \pi d$ or $C = 2\pi r$

		Volume/Capacity	Total Surface Area
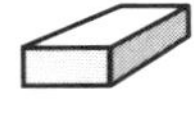	Rectangular Prism	$V = bwh$ or $V = Bh$	$S.A. = 2bh + 2bw + 2hw$ or $S.A. = Ph + 2B$
	Right Circular Cylinder	$V = \pi r^2 h$ or $V = Bh$	$S.A. = 2\pi rh + 2\pi r^2$ or $S.A. = 2\pi rh + 2B$
	Right Square Pyramid	$V = \frac{1}{3}Bh$	$S.A. = \frac{1}{2}P\ell + B$
	Right Circular Cone	$V = \frac{1}{3}\pi r^2 h$ or $V = \frac{1}{3}Bh$	$S.A. = \frac{1}{2}(2\pi r)\ell + B$ or $S.A. = \pi r\ell + \pi r^2$
	Sphere	$V = \frac{4}{3}\pi r^3$	$S.A. = 4\pi r^2$

Sum of the measures of the interior angles of a polygon $= 180(n-2)$

Measure of an interior angle of a regular polygon $= \frac{180(n-2)}{n}$

where:

n represents the number of sides

Algebra 1 and Geometry End-of-Course Assessments Reference Sheet

Slope formula

$$m = \frac{y_2 - y_1}{x_2 - x_1}$$

where m = slope and (x_1, y_1) and (x_2, y_2) are points on the line

Slope-intercept form of a linear equation

$$y = mx + b$$

where m = slope and b = y-intercept

Point-slope form of a linear equation

$$y - y_1 = m(x - x_1)$$

where m = slope and (x_1, y_1) is a point on the line

Distance between two points

$P_1(x_1, y_1)$ and $P_2(x_2, y_2)$

$$\sqrt{(x_2 - x_1)^2 + (y_2 - y_1)^2}$$

Midpoint between two points

$P_1(x_1, y_1)$ and $P_2(x_2, y_2)$

$$\left(\frac{x_1 + x_2}{2}, \frac{y_1 + y_2}{2}\right)$$

Quadratic formula

$$x = \frac{-b \pm \sqrt{b^2 - 4ac}}{2a}$$

where a, b, and c are coefficients in an equation of the form $ax^2 + bx + c = 0$

Special Right Triangles

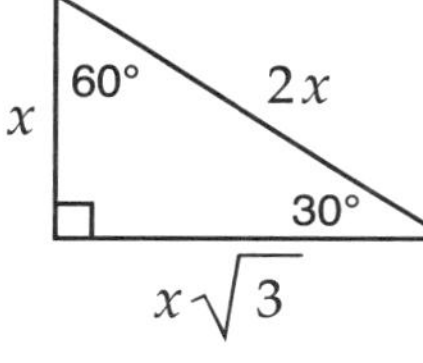

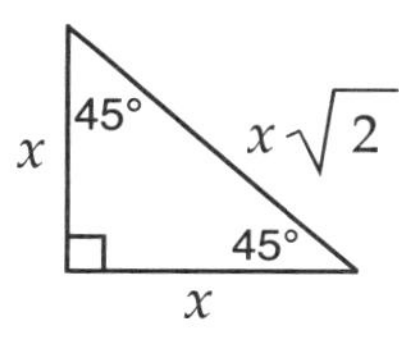

Trigonometric Ratios

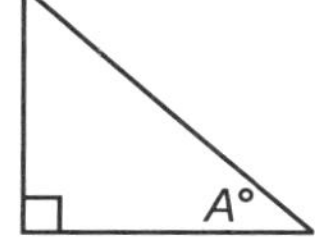

$\sin A° = \dfrac{\text{opposite}}{\text{hypotenuse}}$

$\cos A° = \dfrac{\text{adjacent}}{\text{hypotenuse}}$

$\tan A° = \dfrac{\text{opposite}}{\text{adjacent}}$

Conversions

1 yard = 3 feet
1 mile = 1,760 yards = 5,280 feet
1 acre = 43,560 square feet
1 hour = 60 minutes
1 minute = 60 seconds

1 cup = 8 fluid ounces
1 pint = 2 cups
1 quart = 2 pints
1 gallon = 4 quarts
1 pound = 16 ounces
1 ton = 2,000 pounds

1 meter = 100 centimeters = 1000 millimeters
1 kilometer = 1000 meters
1 liter = 1000 milliliters = 1000 cubic centimeters
1 gram = 1000 milligrams
1 kilogram = 1000 grams

Directions for Taking Geometry EOC Practice Test B

Test Questions

This Practice Test contains 52 questions. The number of questions on the actual test will vary.

- **Multiple-Choice Questions**

  Select the best answer for each question and mark it on the answer sheet on page 314.

- **Open-ended Questions**

  As you come to an open-ended question, use the Notes pages at the back of the book to do your work. Then fill in the answer using the digits 0–9 and/or the symbols for a decimal point, fraction bar, or negative sign in the answer box provided for each specific open-ended question.

Reference Pages

You may refer to the two preceding Reference Pages as often as you like.

Timing

For the actual test you will be given two 80-minute periods to complete the test, with a ten-minute break in between. However, anyone who has not finished will be allowed to continue working.

Checking Your Answers

You will find the correct answers, along with detailed explanations, for this practice test beginning on page 316.

Reviewing Your Work

When finished, turn to the grid on page 326. Circle the number of any questions that you missed in Test B. You will be able to see a pattern that shows which Benchmarks will need your further attention.

1 $\triangle LMN \sim \triangle OPQ$. The area of $\triangle LMN$ is 36 square inches (in^2) and the area of $\triangle OPQ$ is 64 in^2. What is the scale factor of $\triangle LMN$ to $\triangle OPQ$?

A. $\frac{9}{16}$

B. $\frac{3}{4}$

C. $\frac{4}{3}$

D. $\frac{16}{9}$

2 $\triangle ABC \sim \triangle DEF$. Find the length of $\overline{ED}$.

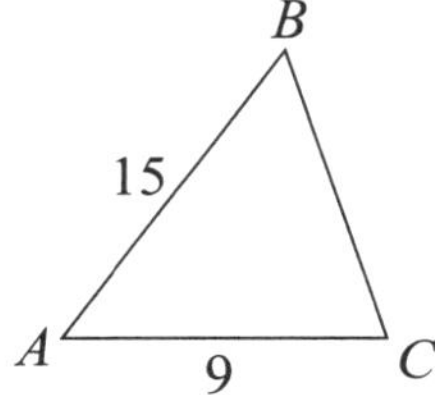

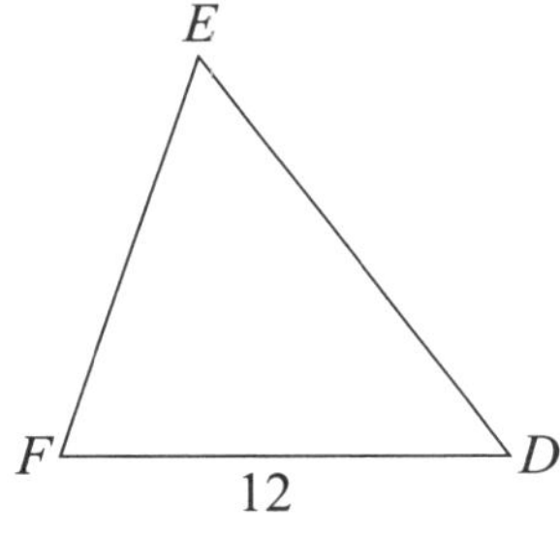

3 Using only the indicated markings, which postulate can be used to justify that the following triangles are congruent?

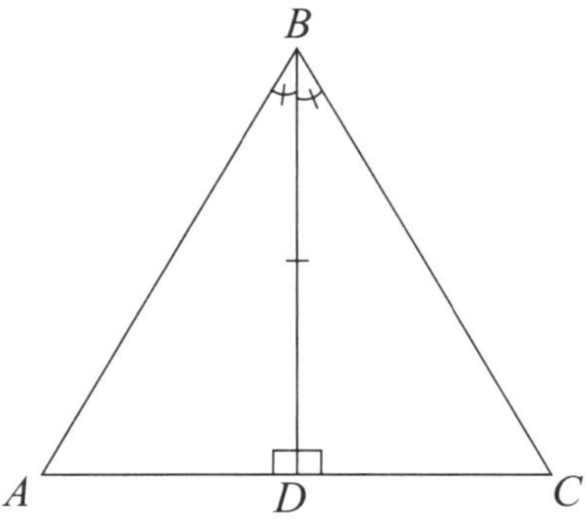

A. SSS

B. SAS

C. ASA

D. AAS

4 Which of the following correctly classifies the triangle below?

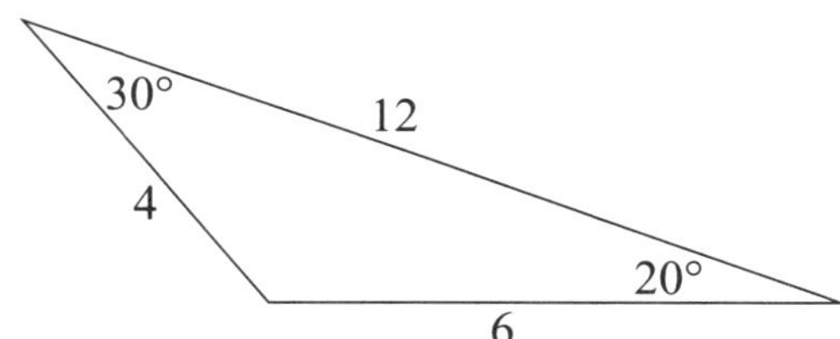

A. isosceles acute

B. isosceles obtuse

C. scalene acute

D. scalene obtuse

5 Which of the following polygons is convex?

A. 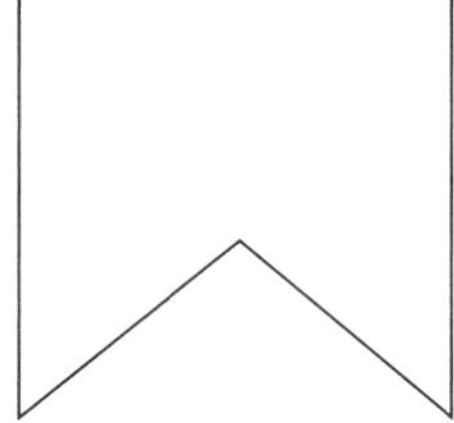

B. 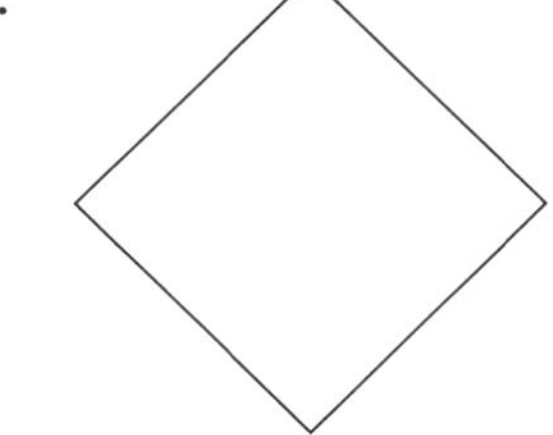

C. 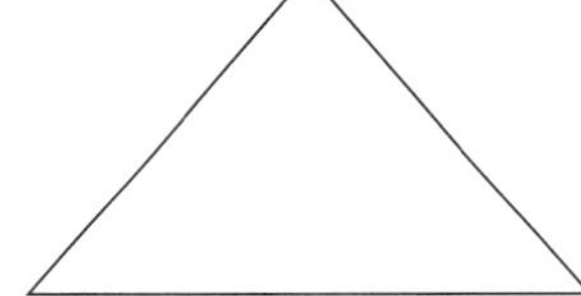

D. both B and C

6 In degrees, what is the sum of the interior angles of a hexagon?

Go On

7. Find the measure of an exterior angle of a regular decagon.

A. 30°

B. 36°

C. 144°

D. 180°

8. Find $m\angle ABC$. *Note*: Figure not drawn to scale.

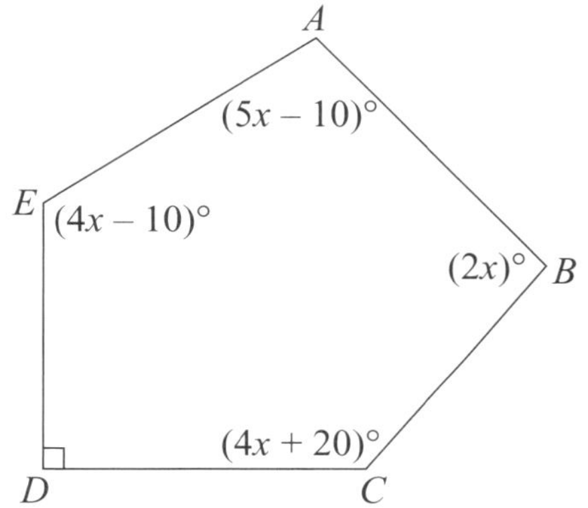

A. 30°

B. 60°

C. 110°

D. 140°

9 $ABCD \sim EFGH$

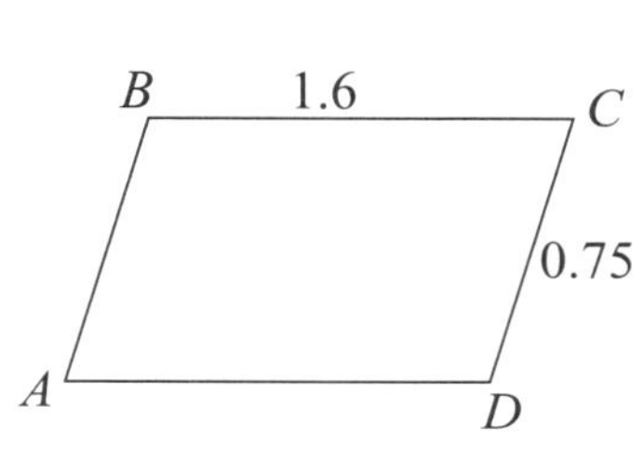

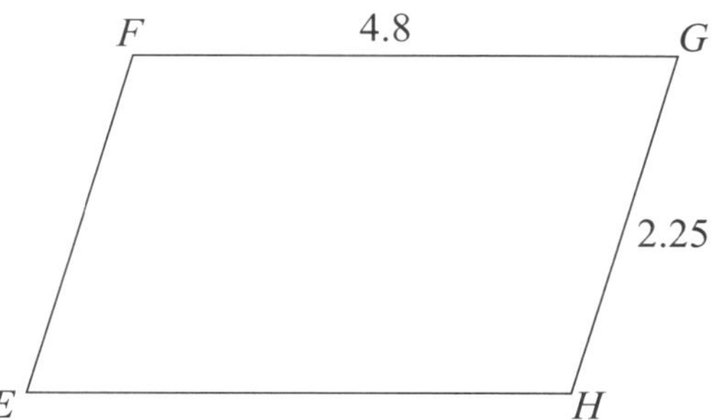

What is the scale factor of $ABCD$ to $EFGH$? Write your answer as a fraction.

10 Given that $\triangle ABC$ and $\triangle LMN$ shown below are congruent, which of the following sides has a length of 12?

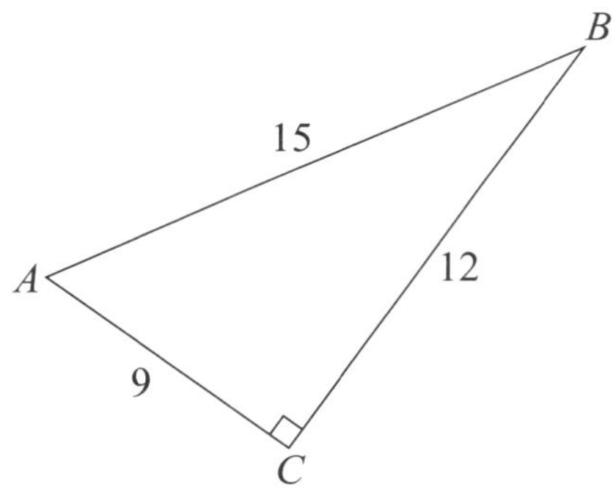

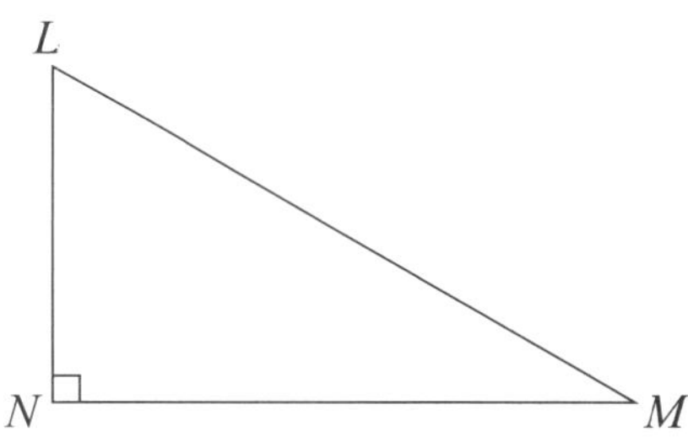

A. $\overline{LN}$

B. $\overline{MN}$

C. $\overline{NL}$

D. None of the above

Use the graph below to answer questions 11 and 12.

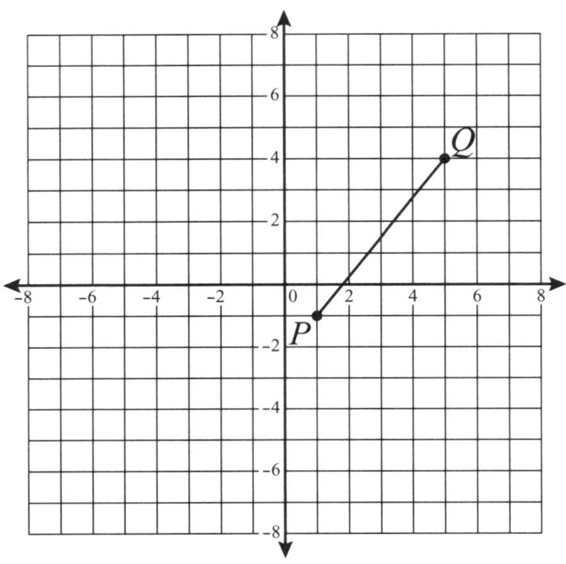

11 What is the length of $\overline{PQ}$?

A. 5

B. $\sqrt{41}$

C. $3\sqrt{5}$

D. $\sqrt{61}$

12 What is the midpoint of $\overline{PQ}$?

A. (2, 1.5)

B. (2, 2.5)

C. (3, 1.5)

D. (3, 2.5)

Go On

13 Which of the following is the value of x?

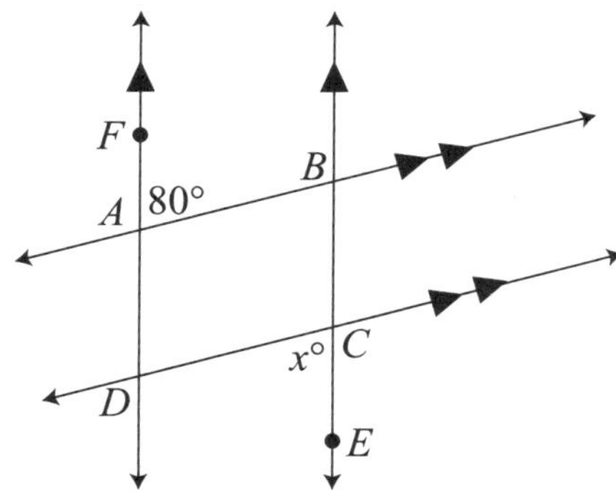

A. 40

B. 80

C. 110

D. 160

14 $\angle 1$ and $\angle 2$ are which type of angle pair?

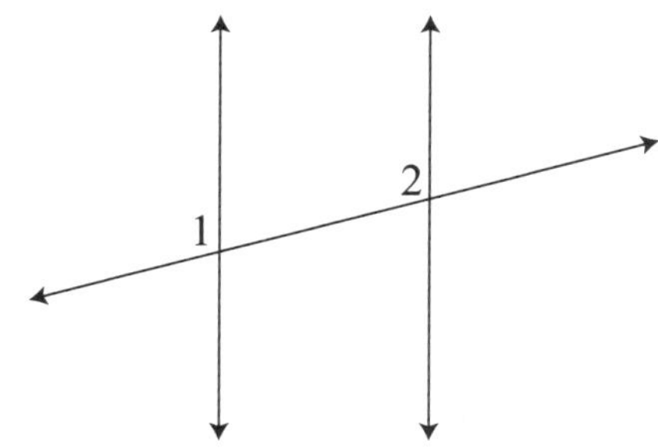

A. corresponding angles

B. same-side interior angles

C. vertical angles

D. alternate interior angles

15 Which of the following is a chord?

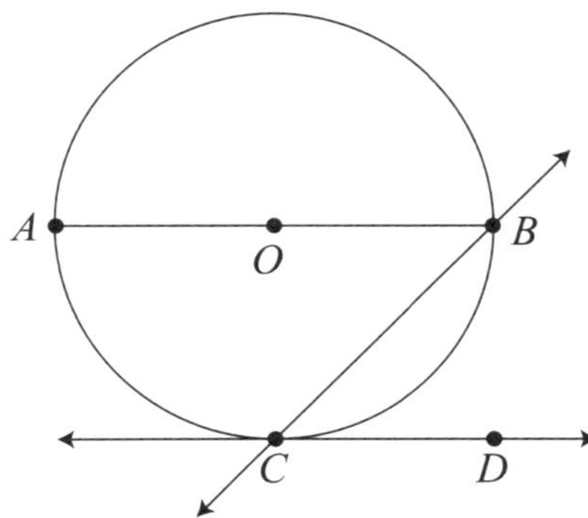

A. $\overline{AO}$

B. $\overline{AB}$

C. $\overleftrightarrow{BC}$

D. $\overleftrightarrow{CD}$

16 Find the length of $\overparen{APB}$ to the nearest tenth of an inch (in.).

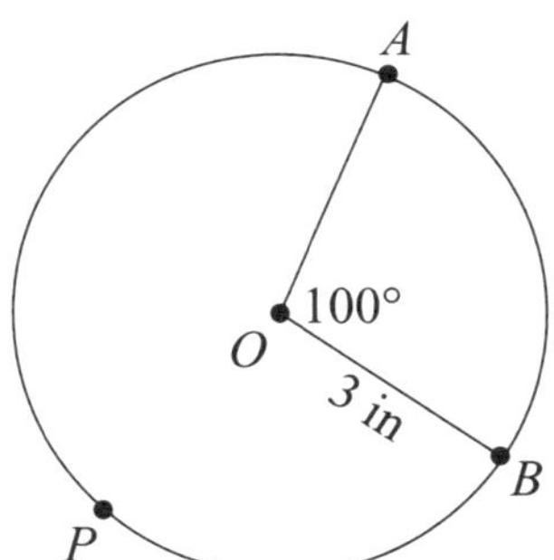

Go On

17 The circumference of a circle is approximately 113.04 centimeters (cm). What is the length of the diameter?

A. 6 cm

B. 12 cm

C. 18 cm

D. 36 cm

18 Braeden lost the lid to his cylindrical garbage can. The top of the garbage can has a radius of 15 inches. Using 3.14 for π, determine the approximate area in square inches of the lid Braeden needs to replace. Round off your answer to the nearest tenth.

19 Find $m\widehat{AC}$.

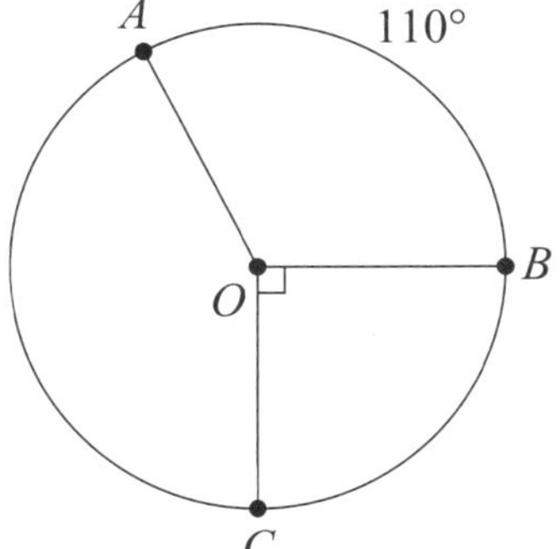

A. 55°

B. 80°

C. 110°

D. 160°

20 Which of the following is the equation of the circle with center (–1, 0) and radius of 4 units?

A. $(x - 1)^2 + y^2 = 4$

B. $(x + 1)^2 + y^2 = 4$

C. $(x - 1)^2 + y^2 = 16$

D. $(x + 1)^2 + y^2 = 16$

21 Which of the following is the graph of $x^2 + (y + 2)^2 = 4$?

A.

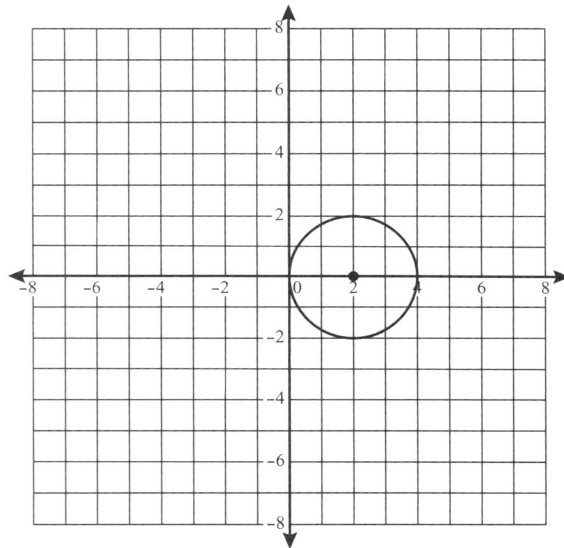

Go On

B.

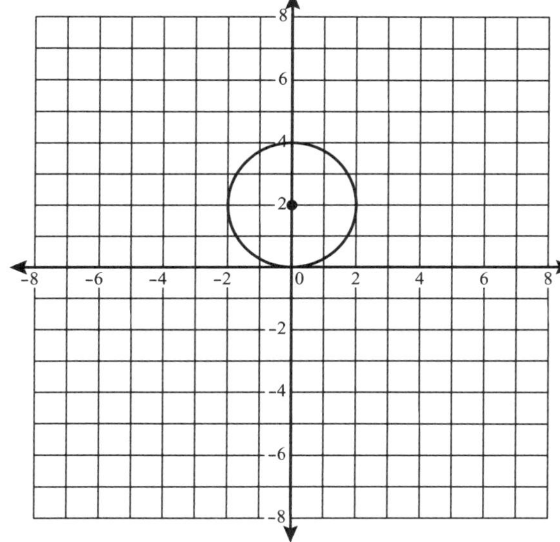

C.

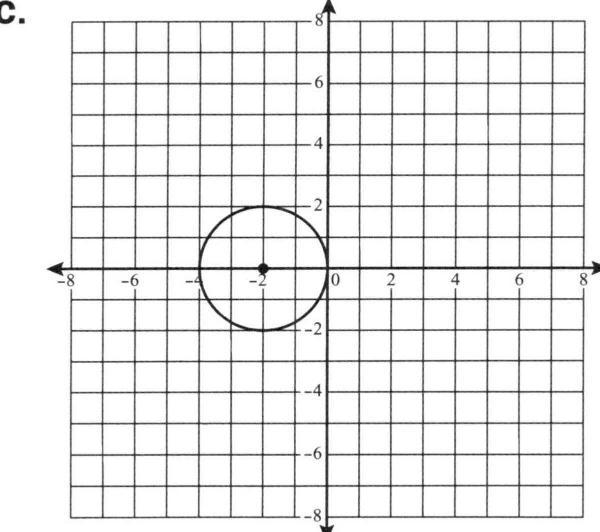

Go On

D.

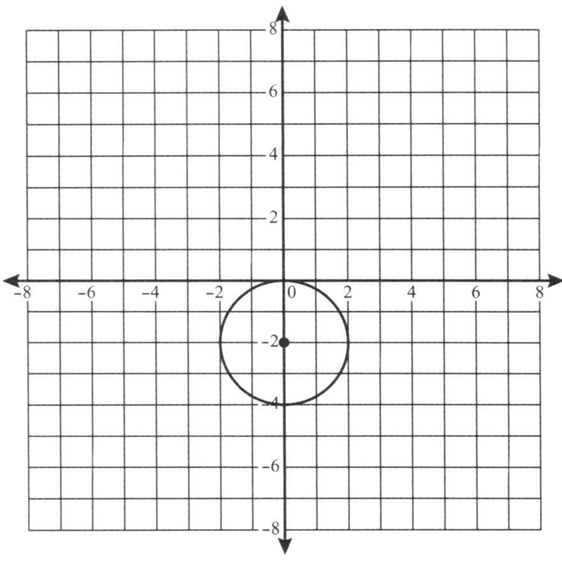

22 The sides of a triangle are 6, 9 and 11. Which of the following correctly classifies the triangle?

A. acute

B. obtuse

C. right

D. equiangular

23 Find the length of $\overline{BC}$ in feet.

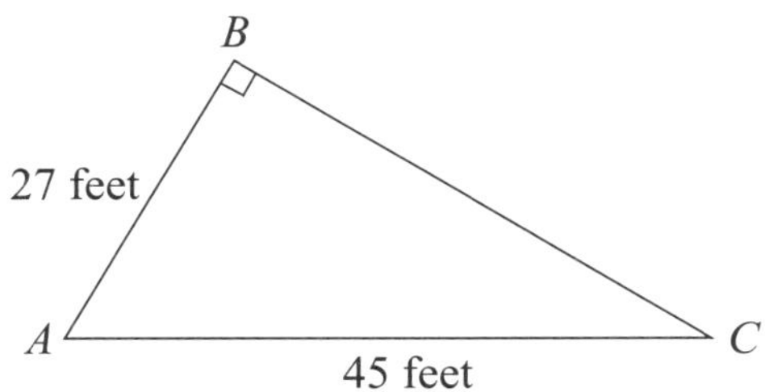

24 The figure below is a net of which solid figure?

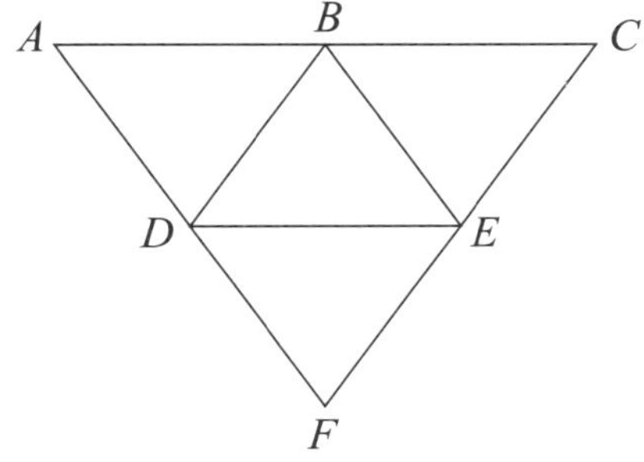

A. pyramid

B. prism

C. cylinder

D. cone

Go On

25 Determine the number of edges in a solid figure that has 12 faces and 10 vertices.

A. 16

B. 18

C. 20

D. 22

26 What type of transformation is shown below?

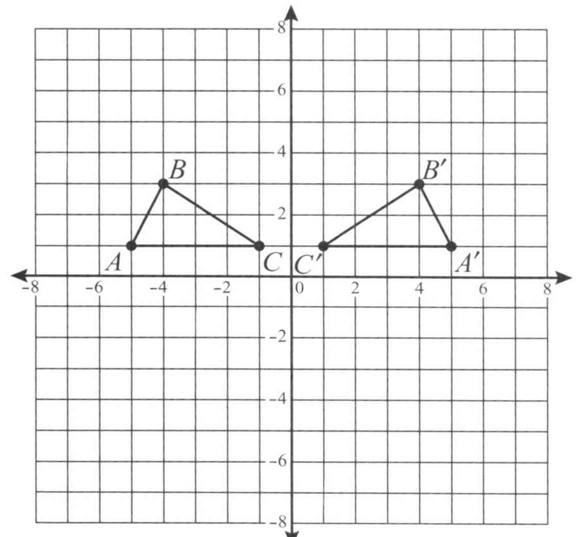

A. reflection

B. rotation

C. translation

D. dilation

Go On

27 Which of the following describes a series of transformations that result in $\triangle A'B'C'$?

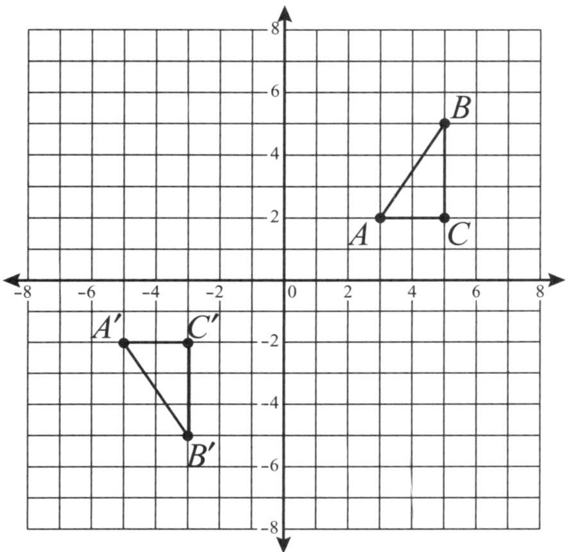

A. reflection over the x-axis, then a reflection of the y-axis

B. reflection over the x-axis, then translate 8 units to the left

C. reflection over the y-axis, then translate 8 units down

D. translate 8 units down, then rotate 90°

28 The area of a trapezoid is 108 square yards. If the bases measure 9 yards and 15 yards, what is the height of the trapezoid in yards?

29 Find the perimeter in inches (in.).

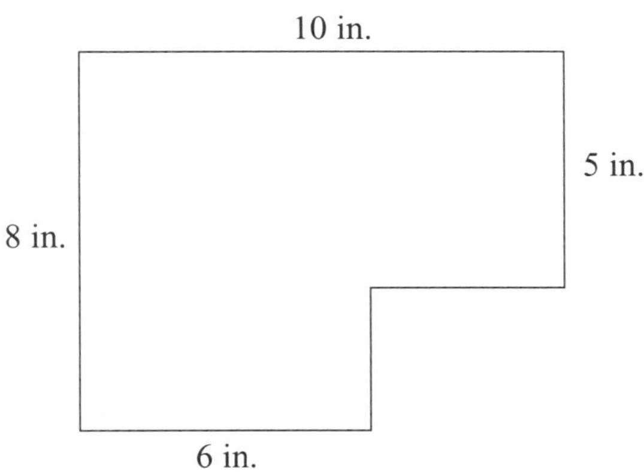

30 Given that the rectangles *ABCD* and *EFGH* are similar, find the area of *EFGH*.

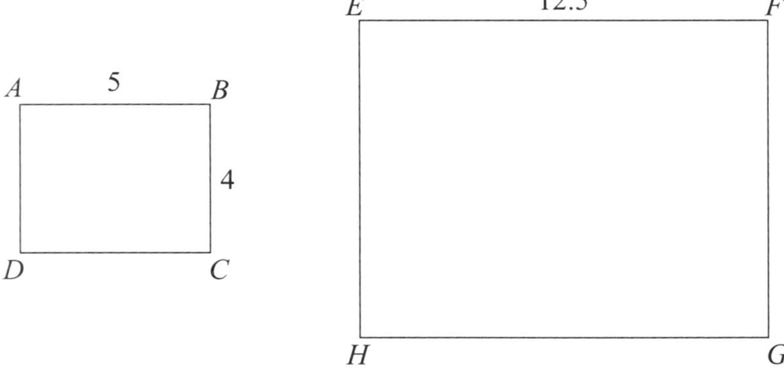

31 Which of the following is a statement that requires a proof?

A. definition

B. postulate

C. theorem

D. known terms

32 Sandy stated that the numerical value of the volume of a cube is always greater than the numerical value of the its surface area. Cynthia found a counterexample by using an edge value of 3. What is the largest integer Cynthia could have selected in order to show that Sandy's statement is incorrect?

33 Thomas stated that two given angles were complementary angles. What did Thomas need to know about the two angles to make this statement?

A. The angles were congruent.

B. The sum of the angles was 90°.

C. The sum of the angles was 180°.

D. The sum of the angles was 360°.

Go On

34 What is the converse of the following statement:

If two lines are perpendicular, then their intersection forms a right angle.

A. If the intersection of two lines forms a right angle, then the lines are perpendicular.

B. If two lines are not perpendicular, then their intersection does not form a right angle.

C. If the intersection of two lines does not form a right angle, then the lines are not perpendicular.

D. If two lines are not perpendicular, then their intersection forms a right angle.

35 Given that the figure below is a rhombus, which statement best explains why the equation $7x + 2 = 5x + 12$ can be used to solve for x?

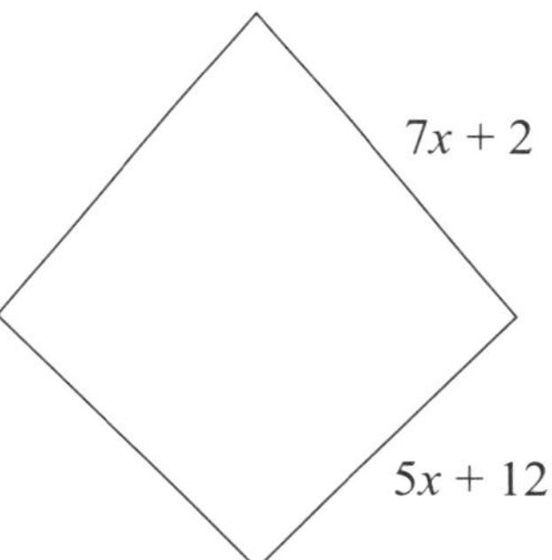

A. A rhombus has two pairs of parallel sides

B. Opposite angles of a rhombus are congruent

C. A rhombus has four congruent sides

D. A rhombus is a type of quadrilateral

36 Which one of the following is logically equivalent to $\sim p \vee q$?

A. $p \rightarrow q$

B. $q \rightarrow p$

C. $p \vee \sim q$

D. $\sim p \wedge q$

37 For which of the following quadrilaterals are the diagonals perpendicular to each other?

A. kite

B. rhombus

C. square

D. all of the above

38 Using the square on the grid below, what is the distance from the midpoint of $\overline{AB}$ to point D? Round off your answer to the nearest hundredth.

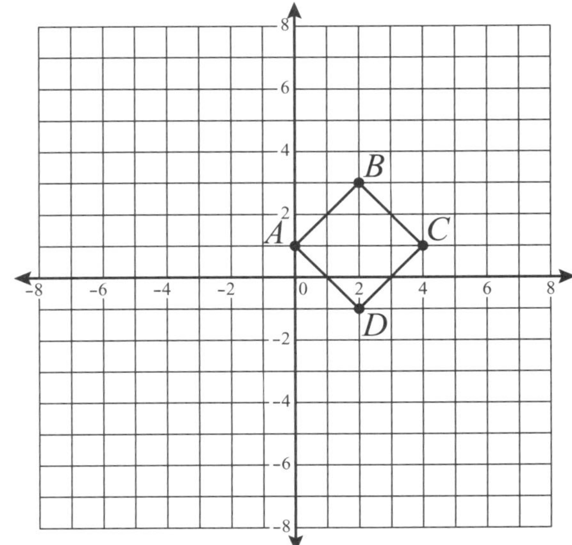

Go On

39 The area of a sphere's great circle is 121π square meters. What is the approximate volume of the sphere in cubic meters?

A. 161π

B. 222π

C. 1775π

D. 14,197π

40 Find the surface area in square feet.

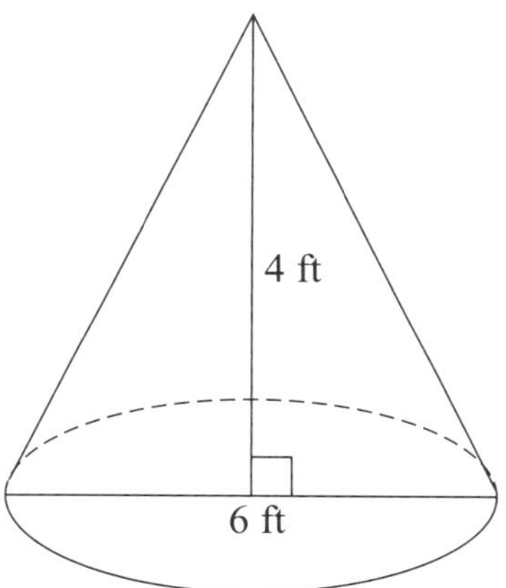

A. 21π

B. 24π

C. 54π

D. 60π

Go On

41 Find the volume in cubic inches (in^3).

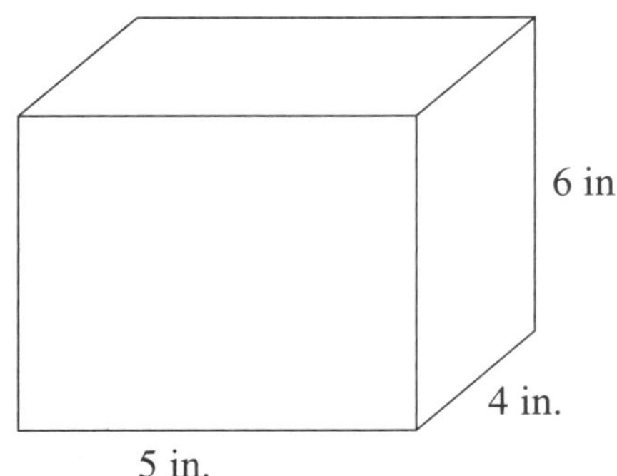

42 Determine the scale factor for the pair of similar solids.

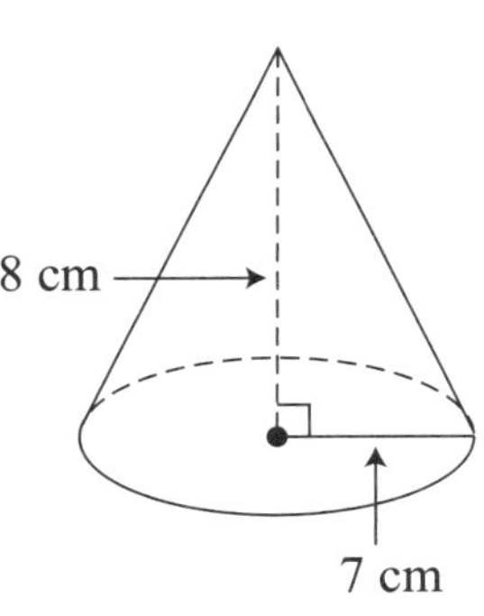

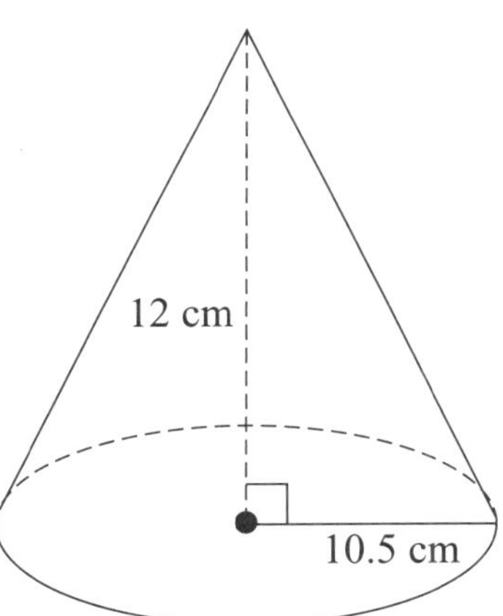

A. $\frac{1}{8}$

B. $\frac{1}{3}$

C. $\frac{2}{3}$

D. $\frac{7}{8}$

Go On

43 Two rectangular prisms are similar. The volume of the smaller prism is 24 cubic inches (in^3) and the volume of the larger prism is 81 in^3. What is the scale factor of these similar prisms? Write your answer as a fraction.

44 Determine the value of x. Round off your answer to the nearest hundredth.

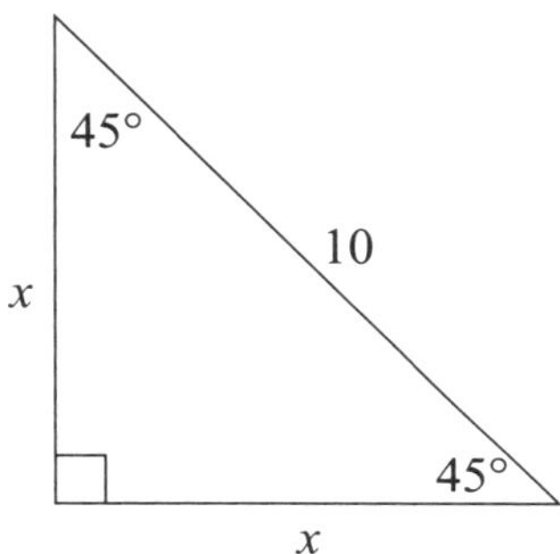

45 Determine the value of x.

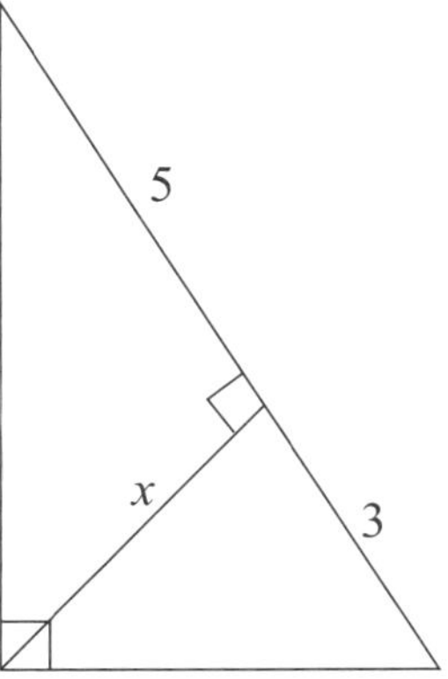

A. $\sqrt{15}$

B. 4

C. $\sqrt{17}$

D. 8

46 Celine has a rectangular piece of wood that measures 10 feet by 15 feet. She is going to use this wood to create a sign. Part of the design is an X that is drawn from opposite corners and is bisected in the center of the wood. She is going to cover the X with ribbon. Approximately how many feet of ribbon will she need to cover the X? Round off your answer to the nearest integer.

Go On

47 Which of the following could be the coordinates of point Y, so that figure $WXYZ$ would be similar to figure $ABCD$?

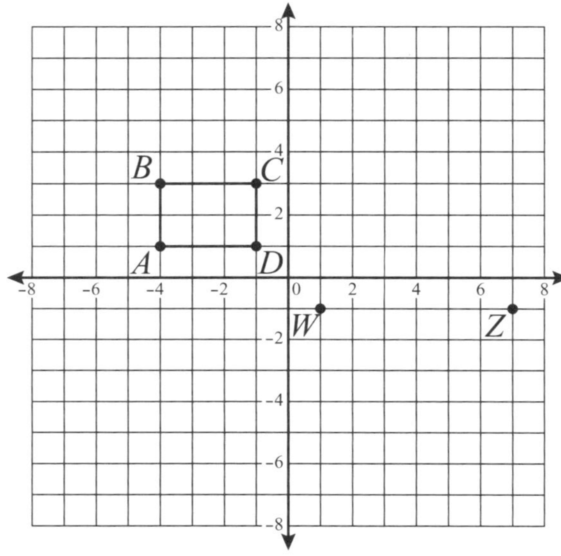

A. (1, –3)

B. (1, 1)

C. (1, 3)

D. (1, 5)

48 A support wire is attached to a 10-foot tree that is perpendicular with the ground. If the wire makes an angle of 55° with the ground, approximately how many feet long is the wire? Round off your answer to the nearest tenth.

49 If P is the centroid of $\triangle ABC$ and $CP = 12$, find the length of DC.

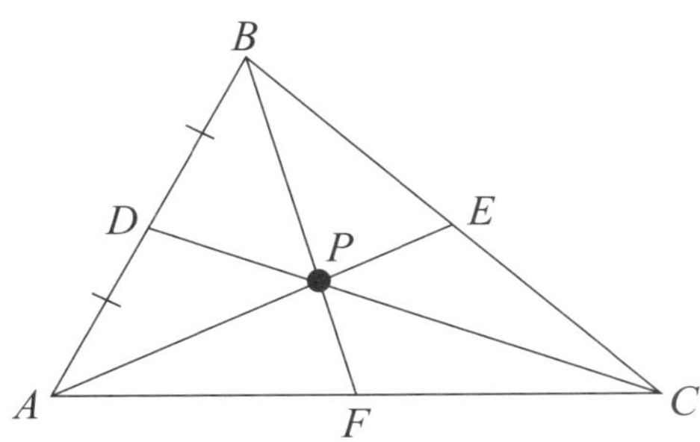

A. 15

B. 16

C. 17

D. 18

50 Find the length of SQ.

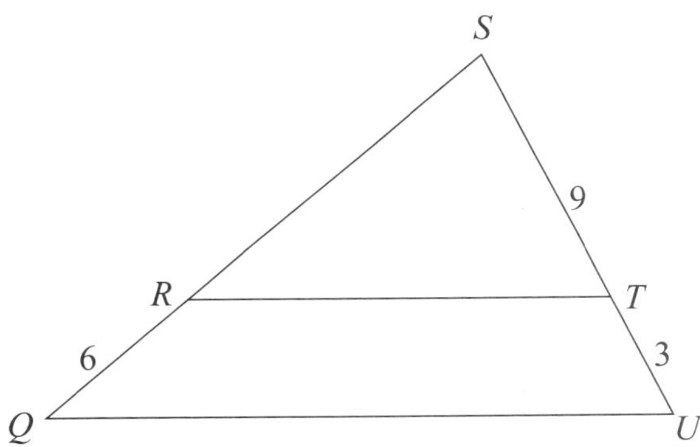

A. 27

B. 24

C. 21

D. 18

51 Which of the following can be the third side of a triangle whose other two sides measure 8 inches and 10 inches?

A. 17 inches

B. 18 inches

C. 19 inches

D. 20 inches

52 In the figure below, $LP = 12$, $LM = 15$, and $PO = 10$. What is the length of $\overline{MO}$ to the nearest hundredth? *Note*: Figure not drawn to scale.

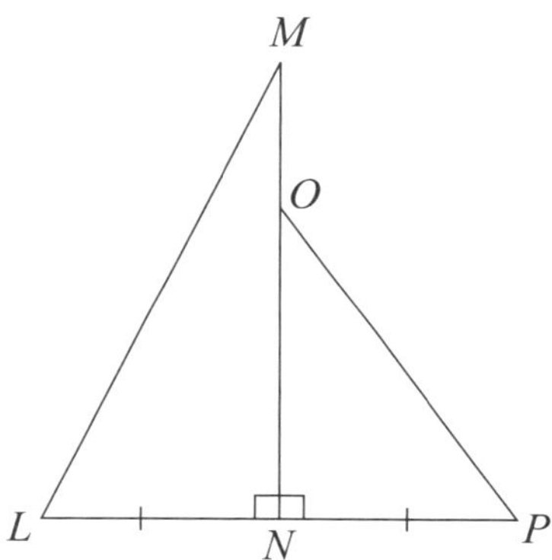

This is the end of the Geometry Test.

Until time is called, go back and check your work or answer questions you did not complete. When you have finished, close your Test Book and Answer Book.

Geometry Practice Test Form B

Answer Sheet

1. Ⓐ Ⓑ Ⓒ Ⓓ
2. ☐☐☐☐☐☐☐
3. Ⓐ Ⓑ Ⓒ Ⓓ
4. Ⓐ Ⓑ Ⓒ Ⓓ
5. Ⓐ Ⓑ Ⓒ Ⓓ
6. ☐☐☐☐☐☐☐
7. Ⓐ Ⓑ Ⓒ Ⓓ
8. Ⓐ Ⓑ Ⓒ Ⓓ
9. ☐☐☐☐☐☐☐
10. Ⓐ Ⓑ Ⓒ Ⓓ
11. Ⓐ Ⓑ Ⓒ Ⓓ
12. Ⓐ Ⓑ Ⓒ Ⓓ
13. Ⓐ Ⓑ Ⓒ Ⓓ
14. Ⓐ Ⓑ Ⓒ Ⓓ
15. Ⓐ Ⓑ Ⓒ Ⓓ
16. ☐☐☐☐☐☐☐
17. Ⓐ Ⓑ Ⓒ Ⓓ
18. ☐☐☐☐☐☐☐
19. Ⓐ Ⓑ Ⓒ Ⓓ
20. Ⓐ Ⓑ Ⓒ Ⓓ
21. Ⓐ Ⓑ Ⓒ Ⓓ
22. Ⓐ Ⓑ Ⓒ Ⓓ
23. ☐☐☐☐☐☐☐
24. Ⓐ Ⓑ Ⓒ Ⓓ
25. Ⓐ Ⓑ Ⓒ Ⓓ
26. Ⓐ Ⓑ Ⓒ Ⓓ
27. Ⓐ Ⓑ Ⓒ Ⓓ
28. ☐☐☐☐☐☐☐
29. ☐☐☐☐☐☐☐
30. ☐☐☐☐☐☐☐
31. Ⓐ Ⓑ Ⓒ Ⓓ
32. ☐☐☐☐☐☐☐
33. Ⓐ Ⓑ Ⓒ Ⓓ
34. Ⓐ Ⓑ Ⓒ Ⓓ
35. Ⓐ Ⓑ Ⓒ Ⓓ
36. Ⓐ Ⓑ Ⓒ Ⓓ
37. Ⓐ Ⓑ Ⓒ Ⓓ
38. ☐☐☐☐☐☐☐
39. Ⓐ Ⓑ Ⓒ Ⓓ
40. Ⓐ Ⓑ Ⓒ Ⓓ
41. ☐☐☐☐☐☐☐
42. Ⓐ Ⓑ Ⓒ Ⓓ
43. ☐☐☐☐☐☐☐
44. ☐☐☐☐☐☐☐
45. Ⓐ Ⓑ Ⓒ Ⓓ
46. ☐☐☐☐☐☐☐
47. Ⓐ Ⓑ Ⓒ Ⓓ
48. ☐☐☐☐☐☐☐
49. Ⓐ Ⓑ Ⓒ Ⓓ
50. Ⓐ Ⓑ Ⓒ Ⓓ
51. Ⓐ Ⓑ Ⓒ Ⓓ
52. ☐☐☐☐☐☐☐

Answers
Practice Test

1 **B**

$$\frac{36}{64} = \frac{a^2}{b^2} \Rightarrow \frac{6}{8} = \frac{a}{b} = \frac{3}{4}$$

2 **20**

$$\frac{9}{12} = \frac{15}{x}$$

$$9x = 180$$

$$x = 20$$

3 **C**

$\angle ABD \cong \angle CBA$

$\overline{BD} \cong \overline{BD}$ by reflexive property

$\angle ADC \cong \angle CDB$

4 **D**

The triangle has three non-congruent angles, which is a scalene triangle. The missing angle is 110°, which is obtuse.

5 **D**

Both B and C are convex because any segment drawn connecting two points in the interior of the polygon lies within the figure.

6 **720**

A hexagon has six sides. With $n = 6$, $(n - 2)180° = (6 - 2)(180)° = 720°$.

7 **B**

The sum of the exterior angles of any polygon is 360°. A regular decagon has 10 congruent exterior angles. Therefore, each angle equals $\frac{360°}{10} = 36°$.

8 **B**

The figure is a pentagon. The sum of the interior angles is 540°. The following equation can be used to solve for x.

$(4x-10)°+(5x-10)°+(2x)°+(4x+20)°+90)°=540°$

$15x+90°=540°$

$15x=450°$

$x=30°$

$m\angle ABC=2x°=2(30)°=60°$

9 $\mathbf{\frac{1}{3}}$

$\frac{1.6}{4.8}=\frac{1}{3}$

$\frac{0.75}{2.25}=\frac{1}{3}$

10 **B**

$\overline{MN}$ corresponds to $\overline{BC}$. Since the triangles are congruent, their corresponding sides are congruent.

11 **B**

$P\,(1,-1)$ and $Q\,(5,4)$

$d=\sqrt{(x_2-x_1)^2+(y_2-y_1)^2}$

$d=\sqrt{(5-1)^2+(4-(-1))^2}=\sqrt{16+25}=\sqrt{41}$

12 **C**

$P\,(1, -1)$ and $Q\,(5, 4)$

$$M = \left(\frac{x_1 + x_2}{2}, \frac{y_1 + y_2}{2}\right)$$

$$M = \left(\frac{1+5}{2}, \frac{-1+4}{2}\right) = (3,\ 1.5)$$

13 **B**

$\angle ADC \cong \angle FAB$, since they are corresponding angles of parallel lines. So, $m\angle ADC = 80°$. Also, $\angle ADC \cong \angle DCE$, since they are alternate interior angles of parallel lines. So, $x = 80$.

14 **A**

By definition, two angles on the same side of a transversal and in the same relative position are corresponding angles.

15 **B**

$\overline{AB}$ is a chord because its endpoints are on the circle.

16 **13.6**

length $\overparen{APB} = \frac{260}{360} \cdot 2\pi(3) \approx 13.6$.

17 **D**

$C = \pi d$

$$\frac{113.04}{\pi} = \frac{\pi d}{\pi}$$

Use 3.14 for the value of π.

$36 = d$

18 **706.5**

$A = \pi r^2 = \pi(15)^2 \approx 706.5$

19 **D**

$m\overset{\frown}{AC} + m\overset{\frown}{CB} + m\overset{\frown}{AB} = 360°$

$m\overset{\frown}{AC} + 90° + 110° = 360°$

$m\overset{\frown}{AC} + 200° = 360°$

$m\overset{\frown}{AC} = 160°$

20 **D**

The equation of a circle with center (h, k) and radius r is $(x - h)^2 + (y - k)^2 = r^2$. By substitution, the required equation is $(x - (-1))^2 + (y - 0)^2 = 4^2 \Rightarrow (x + 1)^2 + y^2 = 16$.

21 **D**

Rewrite the given equation as $(x - 0)^2 + (y - (-2))^2 = 2^2$. Then the center of the circle is $(0, -2)$ and the radius is 2.

22 **B**

$11^2 \underline{\ ?\ } 6^2 + 9^2$

$121 \underline{\qquad} 36 + 81$

$121 > 117$ obtuse

23 **36**

$a^2 + b^2 = c^2$

$27^2 + b^2 = 45^2$

$729 + b^2 = 2025$

$b^2 = 1296$

$b = 36$

24 **A**

As a three-dimensional figure, the base will be $\triangle BDE$ and the triangular faces will be *ABD*, *CBE*, and *DEF*. By definition, this is a pyramid (with a triangular base).

25 **C**

$V - E + F = 2$

$10 - E + 12 = 2$

$22 - E = 2$

$-E = -20$

$E = 20$

26 **A**

$\triangle ABC$ is reflected over the *y*-axis. This means that the point (x, y) becomes $(-x, y)$. As an example, point *A* located at $(-5, 1)$ becomes point A' located at $(5, 1)$. Each of points *B* and *C* is affected in the same way.

27 **B**

When $\triangle ABC$ is reflected over the *x*-axis, the point (x, y) becomes $(x, -y)$. After a translation of 8 units left, $(x, -y)$ becomes $(x - 8, -y)$. As an example, point *A* located at $(3, 2)$ becomes point A' located at $(-5, -2)$. Each of points *B* and *C* is affected similarly.

28 **9**

$$A = \frac{1}{2}h(b_1 + b_2)$$

$$108 = \frac{1}{2}h(9 + 15)$$

$$108 = \frac{1}{2}h(24)$$

$$108 = 12h$$

$$9 = h$$

29 **36**

8 in. + 10 in. + 5 in. + 4 in. + 3 in. + 6 in. = 36 in.

30 **125**

The scale factor is $\frac{5}{12.5} = \frac{2}{5}$. The ratio of the areas is $\frac{2^2}{5^2} = \frac{4}{25}$. The area of *ABCD* is 20, so $\frac{4}{25} = \frac{20}{x}$ and $x = 125$.

31 **C**

Theorems always require proofs.

32 **6**

The volume of a cube with an edge of x is x^3. The surface area of a cube with an edge of x is $6x^2$. In order to disprove Sandy's claim, we need $6x^2 \geq x^3$. Divide both sides by x^2 to get $6 \geq x$. Thus, 6 is the largest allowable integer. (Note that if $x = 6$, the numerical value of both the surface area and volume is 216.)

33 **B**

By definition, complementary angles have a sum of 90°. They need not be congruent.

34 **A**

The converse switches the hypothesis and the conclusion.

35 **C**

By definition, a rhombus has four congruent sides.

36 **A**

Here are the truth tables for $\sim p \vee q$ and $p \rightarrow q$.

p	q	$\sim p$	$\sim p \vee q$
T	T	F	T
T	F	F	F
F	T	T	T
F	F	T	T

p	q	$p \rightarrow q$
T	T	T
T	F	F
F	T	T
F	F	T

Thus, $\sim p \vee q$ is logically equivalent to $p \rightarrow q$.

37 **D**

For each of these quadrilaterals, the diagonals are perpendicular to each other, as shown below.

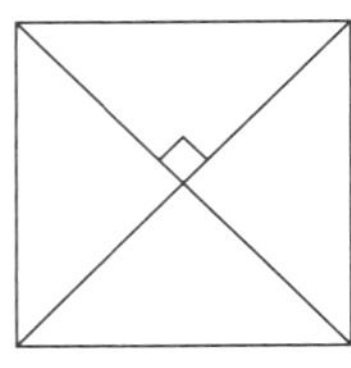

Square

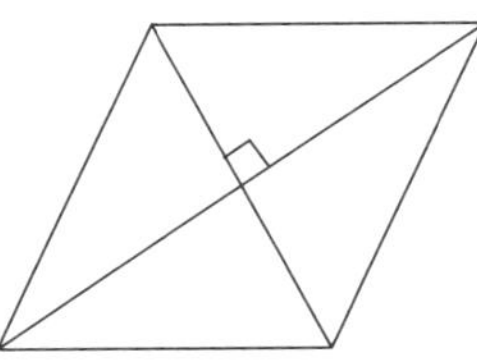

Rhombus

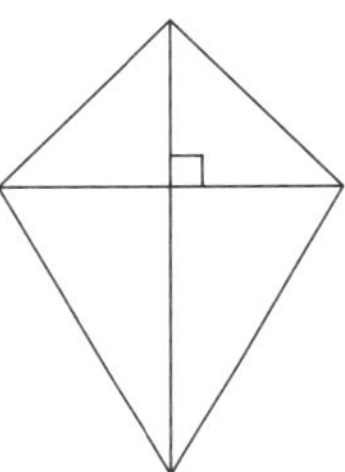

Kite

38 **3.16**

The midpoint of $\overline{AB}$ is (1, 2). Then the distance between (1, 2) and (2, –1) is $\sqrt{(1-2)^2+(2-(-1))^2} = \sqrt{1+9} = \sqrt{10} \approx 3.16$.

39 **C**

To find the radius of the sphere, use the area of the great circle to solve for r.

$A = \pi r^2$

$121\pi = \pi r^2$

$121 = r^2$

$11 = r$

Volume of a sphere $= \frac{4}{3}\pi r^3 = \frac{4}{3}\pi(11)^3 = \frac{4}{3}\pi(1331) \approx 1775\pi$.

40 **B**

We need to find the slant height ℓ. Using the Pythagorean theorem, $4^2 + 3^2 = \ell^2$. (Remember that 6 ft is the diameter, not the radius of the circle.) Then $25 = \ell^2$, so $\ell = 5$.

Surface area $= \pi r^2 + \pi r\ell = (\pi)(3^2) + \pi(3)(5) = 24\pi$.

41 **120**

$V = Bh = (5)(4)(6) = 120$

42 **C**

$\frac{8}{12} = \frac{2}{3}$

$\frac{7}{10.5} = \frac{2}{3}$

43 $\mathbf{\frac{2}{3}}$

$\frac{a^3}{b^3} = \frac{24}{81} = \frac{8}{27}$

$\frac{a^3}{b^3} = \frac{8}{27} \Rightarrow \frac{\sqrt[3]{a^3}}{\sqrt[3]{b^3}} = \frac{\sqrt[3]{8}}{\sqrt[3]{27}} = \frac{2}{3}$

44 **7.07**

$$x\sqrt{2} = 10 \Rightarrow \frac{x\sqrt{2}}{\sqrt{2}} = \frac{10}{\sqrt{2}} \Rightarrow x = \frac{10}{\sqrt{2}}$$

Rationalize the denominator: $\frac{10}{\sqrt{2}} \cdot \frac{\sqrt{2}}{\sqrt{2}} = \frac{10\sqrt{2}}{2} = 5\sqrt{2} \approx 7.07.$

45 **A**

$$\frac{5}{x} = \frac{x}{3} \Rightarrow x^2 = 15 \Rightarrow x = \sqrt{15}$$

46 **36**

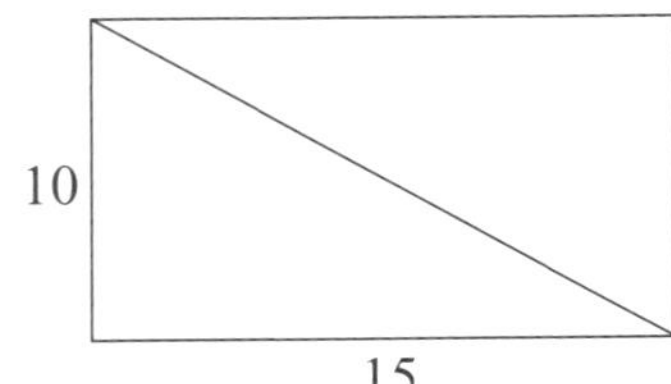

$$10^2 + 15^2 = c^2$$

$$100 + 225 = c^2$$

$$325 = c^2$$

$$18.1 \approx c$$

Since there are two diagonals, the length of one diagonal needs to be doubled: $(18.1)(2) = 36.2 \approx 36$.

47 **C**

If the length of the rectangle is doubled from 3 units to 6 units, the width must be doubled from 2 units to 4 units.

48 **12.2**

Let x represent the length of the wire.

$$\sin 55° = \frac{10}{x}$$

$$x(\sin 55°) = 10$$

$$x(.82) = 10$$

$$x \approx 12.2$$

49 **D**

$$CP = \frac{2}{3}DC$$

$$12 = \frac{2}{3}DC$$

$$18 = DC$$

50 **B**

Let $SR = x$

$$\frac{x}{6} = \frac{9}{3}$$

$$3x = 54$$

$$x = 18$$

Then $SQ = RQ + SR = 6 + 18 = 24$.

51 **A**

The sum of 8 and 10 must be greater than the third side. The sum of 8 and 10 is 18, which is greater than 17.

52 **5.75**

$LN = NP = \left(\frac{1}{2}\right)(LP) = 6$. In $\triangle LMN$, $6^2 + (MN)^2 = 15^2$. Then $(MN)^2 = 15^2 - 6^2 = 225 - 36 = 189$. So $MN = \sqrt{189} \approx 13.75$. In $\triangle MNP$, $6^2 + (ON)^2 = 10^2$. Then $(ON)^2 = 10^2 - 6^2 = 100 - 36 = 64$. So, $ON = \sqrt{64} = 8$. Thus, $MO = MN - ON \approx 5.75$.

STANDARDS FOR FLORIDA GEOMETRY END-OF-COURSE

STANDARD	PRACTICE TEST A	PRACTICE TEST B
MA.912.G.1.1	10, 11	11, 12
MA.912.G.1.3	12	13, 14
MA.912.G.2.1	51	5
MA.912.G.2.2	7	6, 7, 8
MA.912.G.2.3	8, 9	9, 10
MA.912.G.2.4	23, 24	26, 27
MA.912.G.2.5	25, 26	28, 29
MA.912.G.2.7	37	30
MA.912.G.3.1	34	37
MA.912.G.3.2	34	37
MA.912.G.3.3	44	47
MA.912.G.3.4	32, 35	35, 38
MA.912.G.4.1	6	4
MA.912.G.4.2	46	49
MA.912.G.4.4	3, 4	1, 2
MA.912.G.4.5	49	50
MA.912.G.4.6	5	3
MA.912.G.4.7	48, 50	51
MA.912.G.5.1	1	22, 23
MA.912.G.5.2	2	45
MA.912.G.5.3	42, 43	44
MA.912.G.5.4	1, 20	46
MA.912.G.6.2	13	15, 16, 17
MA.912.G.6.4	16, 17	19
MA.912.G.6.5	14, 15	18
MA.912.G.6.6	18	20
MA.912.G.6.7	19	21
MA.912.G.7.1	21	24
MA.912.G.7.2	22	25
MA.912.G.7.4	36	39
MA.912.G.7.5	38, 39	32, 40, 41
MA.912.G.7.6	40	42
MA.912.G.7.7	41	43
MA.912.G.8.1	27	31
MA.912.G.8.2	8, 25	46
MA.912.G.8.3	28	32
MA.912.G.8.4	30	33
MA.912.G.8.5	47	52
MA.912.T.2.1	29, 45	48
MA.912.D.6.2	31	34
MA.912.D.6.3	33	36
MA.912.D.6.4	47	52